A MODERN HANDRAIL

Common-Sense Stair Building *and* Handrailing

Handrailing in Three Divisions
SHOWING THREE OF THE SIMPLEST METHODS KNOWN IN THE ART, WITH COMPLETE INSTRUCTIONS FOR LAYING OUT AND WORKING HANDRAILS SUITABLE FOR ANY KIND OF A STAIR, STRAIGHT, CIRCULAR OR ELLIPTICAL, OR FOR STAIRS WITH LANDINGS AND CYLINDERS

Stair Building
COVERS UPWARDS OF EIGHTY PAGES, DEVOTED TO NEWEL OR PLATFORM STAIRS CHIEFLY, GIVING INSTRUCTIONS FOR THEIR BUILDING, PLANNING AND DECORATION

BY

FRED T. HODGSON, *Architect*

ILLUSTRATED WITH OVER TWO HUNDRED AND FIFTY DRAWINGS AND DIAGRAMS, AND CONTAINING A GLOSSARY OF TERMS USED IN STAIR BUILDING AND HANDRAILING, AND IN ADDITION, TWENTY-FIVE MODERATE PRICED HOUSE DESIGNS, SHOWING THE PERSPECTIVE VIEW AND FLOOR PLANS.

Fredonia Books
Amsterdam, The Netherlands

Common-Sense Stair Building and Handrailing

by
Fred T. Hodgson

ISBN: 1-4101-0167-3

Copyright © 2003 by Fredonia Books

Reprinted from the 1963 edition

Fredonia Books
Amsterdam, The Netherlands
http://www.fredoniabooks.com

All rights reserved, including the right to reproduce this book, or portions thereof, in any form.

PREFACE

The following pages in stair-building and handrailing are taken from the actual working drawings of practical handrailers and stair-builders. The first division is, in a great measure, the work of George Langstaff, New England, and is considered by expert workmen to be one of the best treatises of the kind, with regard to the stairs dealt with. Of course there are only eleven kinds of stairs, but they are so arranged that any person mastering to the full extent these eleven would find no insurmountable difficulty in dealing with stairs of other kinds.

It must be remembered that the reader of this book is supposed to have a considerable knowledge regarding the various methods of building the stair proper in all its different forms, for without this knowledge it will be impossible to understand the method of laying out and constructing a rail, even for a straight stair having a ramp at the newel post. That is the publishers' reason for including a valuable treatise on that subject, which teaches, in a very simple manner, the proper way to lay out the carcass of a stair, and all new beginners who have not obtained a fair knowledge on the subject will appreciate this addition, which, in conjunction with this work, will fully equip any young man with all the information he will ever likely require regarding the art of stair-building and handrailing.

PREFACE

The greater portion of the first division was published in "The Builder and Woodworker" many years ago, and afterwards, in a very much amended form, in "The National Builder," and is now in book form for the first time.

The second division which contains some excellent examples is the work of several contributors, who worked under a like system. The methods of obtaining the wreaths and twists are worth studying, as they show how these can be lined out with the greatest of ease when the subject is understood. This method is nearly complete in itself.

The third division is perhaps the most complete of the three, as about any kind of a rail can be obtained by the use of this system. While not exactly like the system of the late Robert Riddell, it approaches it so nearly that ordinary workmen would scarcely know the difference, but there is a difference, and Mr. Wilson, who has helped to work this system out, deserves much credit for simplifying the whole scheme.

The science of handrailing was never reduced to such simplicity as now, and it is claimed for the three divisions shown in "Common-Sense Handrailing" that the latest and simplest methods are shown therein, and this, too, at about one-fifth the cost of the older and more elaborate methods. In saying this we do not mean to belittle the larger and in some cases the more extended works of Nicholson, Graff, Reynalds, Sherrett. Monckton. Secor. Riddell and others. Each

PREFACE

and every one has much to recommend it, and the expert handrailer will no doubt have copies of these larger works on his shelves. To the first and last of the names given in the foregoing belong the greatest honors in this science, the first for his invention, or rather discovery, of the true geometrical principles involved, and the latter for divesting the science of its crudities and reducing it to more simple conditions. Nearly all improvements in the science are due in large measure to the methods employed by Robert Riddell.

The prismatic solid when thoroughly understood will show to the student pretty nearly everything required in handrailing, and it is the advice of the writer that this solid should be analyzed by the young man who wishes to become an expert, and the study will neither be tedious nor uninteresting.

In all cases a stairway should be commodious, inviting and easy of ascent, and when winders are used they should extend past the spring line of the cylinder, so as to give a fair wreath at narrow end of tread and to bring the rail as near as possible to the same pitch as rail over square steps, and when the hall or space is sufficiently wide should not be less than 3 feet 6 inches in width; 4 feet would be much better, then two persons can pass each other. The height of riser and width of step are governed by the space allowed for the stairs, but as a general rule the step should not be less than 9 inches wide and the riser should not exceed

8 inches in height. Seven inches rise and 11 inches tread make a very easy and good-looking stairway. If the width of tread is increased the riser must be correspondingly reduced. The tread and riser together should not be over 18 inches or less than 17 inches. Of course there are occasions when this rule cannot be employed, and the workman will be called upon to exercise his own judgment, but the closer he keeps to this rule the better will be his stair so far as comfort and convenience are concerned.

This little book contains over 240 illustrations—all of a practical nature—and it is hoped the text describing them is sufficiently clear, and that the student will have no difficulty in understanding what is meant and in being able, after understanding them, to construct a handrail over any flight of stairs that he may be called upon to erect. This is the ardent wish of the writer.

FRED T. HODGSON.

January, 1903.

Common-Sense Handrailing

FIRST METHOD

The building of stairs and properly making and placing over them a graceful handrail and suitable balusters and newel posts is one of the greatest achievements of the joiner's art and skill, yet it is an art that is the least understood of any of the constructive processes, that the carpenter or joiner is called upon to accomplish. In but very few of the plans made by an architect are the stairs properly laid down or divided off; indeed, most of the stairs as laid out and planned by the architect, are impossible ones owing to the fact that the circumstances that govern the formation of the rail are either not understood, or not noticed by the designor; and the expert handrailer often finds it difficult to conform the stairs and rail to the plan. Generally, however, he gets so close to it that the character of the design is seldom changed.

The stairs are the great feature of a building, as they are the first object that meets the visitor and claims his attention, and it is essential, therefore, that the stair and its adjuncts should have a neat and graceful appearance, and this can only be accomplished by having the rail properly made and set up.

It is proposed in this little book to give such instructions in the art of handrailing as will enable the young workman to build a rail so that it will assume a handsome appearance when set in place. There are eleven distinct styles of stairs shown, but the same principle that governs the making of the simplest rail, governs the construction of the most difficult, so, having once mastered the simple problems in this system, progress in the art will become easy, and a little study and practice will enable the workman to construct a rail for the most tortuous stairway.

A knowledge of geometry is not required in the study of this system, but it would aid the workman materially if he possessed a knowledge of that science, and where possible he should avail himself of acquiring as much knowledge of geometry as possible, not only for the study of handrailing but nearly every branch of the building trade.

The progressive lessons given herewith will, I am sure, be of great assistance to stair-builders already engaged in the business and to the young aspiring mechanic, anxious to master every branch of his trade and to penetrate all its mysteries. This system will open a hitherto sealed book, especially to the young man whose knowledge of geometry may be rather limited. There will be no labyrinthic network of lines to torment and confuse the student, nothing but what is absolutely necessary to obtain the face moulds and bevels for marking and working the wreaths. The figures from 1 to 11 show flights of stairs of various shapes and forms, and cover all the examples the workman will ever likely be called upon to build. At any rate, if he should have to construct a form of stairs not shown in these examples, the knowledge gained by a study of these presented will enable him

FIRST METHOD

to wrestle with other forms, no matter what their plans may be. The only form of stair not shown that the student may be called upon to build would very likely be flights having an elliptical plan, but as this form is so seldom used, and then only in public buildings or great mansions, it seldom falls to the lot of the ordinary workman to be called upon to design or construct them. However, to provide for such a contingency a method of laying out and constructing a handrail will be illustrated and described at the close of this treatise.

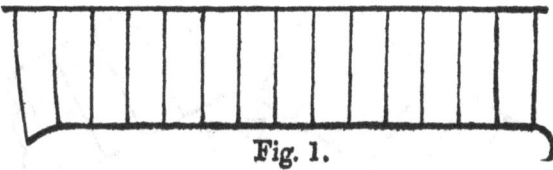

Fig. 1.

Fig. 1 exhibits the plan of a straight stair with an ordinary cylinder at the top, provided for a return rail on the landing. It also shows a lengthened step at the starting.

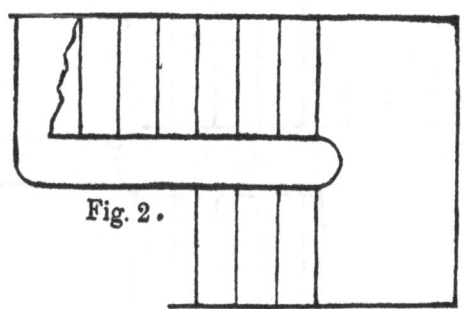

Fig. 2.

Fig. 2 shows a plan of a stair with a landing and return steps.

COMMON-SENSE HANDRAILING

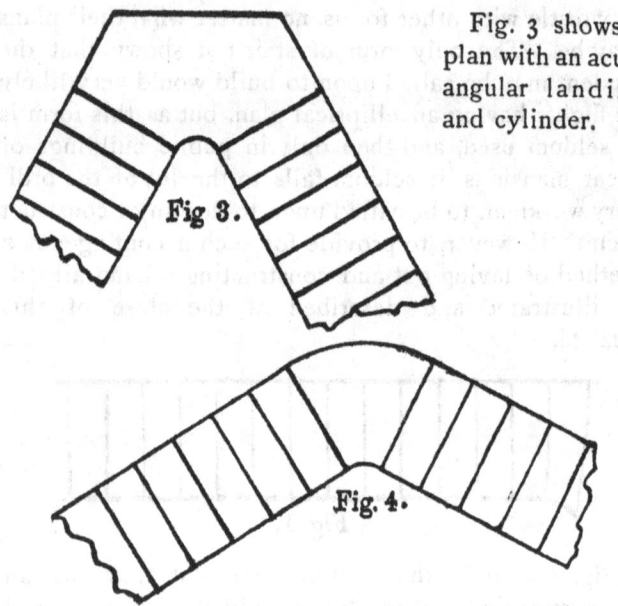

Fig. 3 shows a plan with an acute angular landing and cylinder.

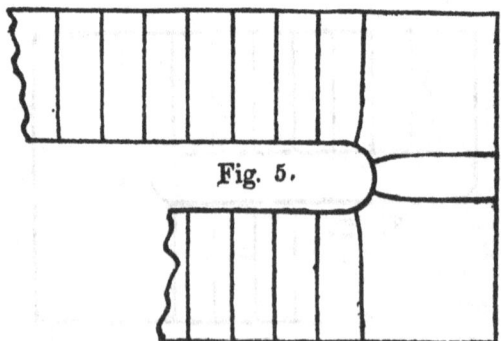

Fig. 4 shows the same kind of stair as Fig. 3, only being at an obtuse angle.

Fig. 5 exhibits a stair having a half-turn with two risers on landings.

FIRST METHOD

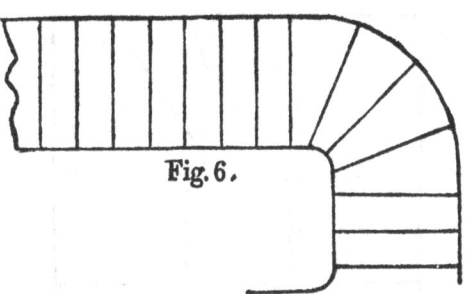

Fig. 6 shows a plan of a quarter-space stair with four winders.

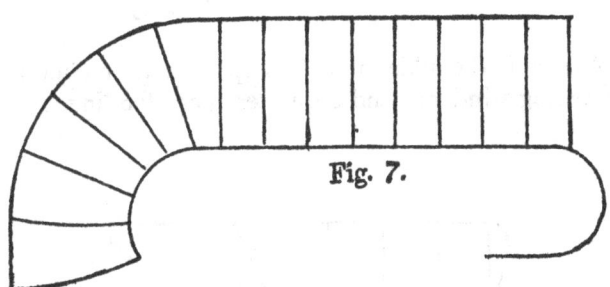

Fig. 7 is the plan of a stair similar to Fig. 6, but having seven winders.

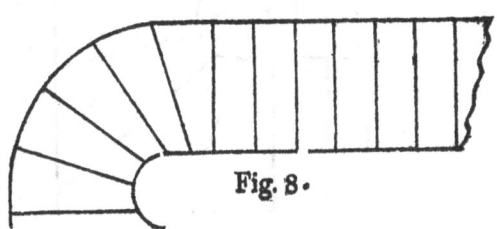

Fig. 8 shows the plan of a stair having five "dancing winders."

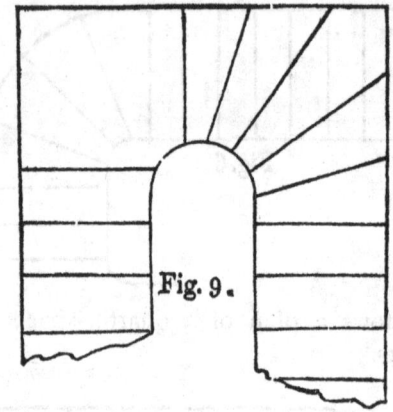

Fig. 9 is the plan of a half-space stair having five "dancing winders" and a quarter-space landing.

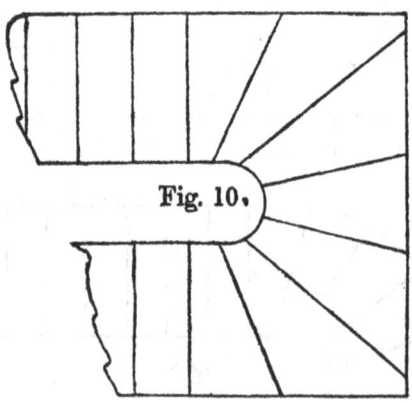

Fig. 10 shows the plan of a half-space stair with 'dancing winders" all around the cylinder

FIRST METHOD

Fig. 11 shows the plan of a geometrical stair having winders all around the cylinder.

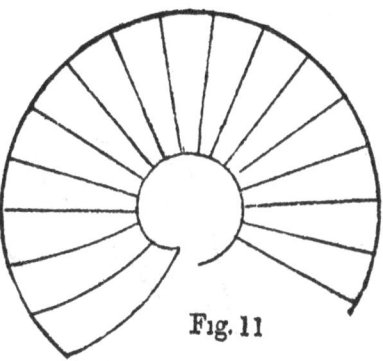

Fig. 11

As it is necessary the student should be acquainted with the methods of development of the angle of tangents which give shape and joints of the face moulds directly from the pitch lines, a couple of examples are herewith illustrated. Fig. 12 shows a straight pitch in which both tangents are of equal length, while Fig. 13 shows the tangents of unequal lengths and different pitches, and I advise the student to thoroughly master these two problems by frequently reproducing them, as these two examples are the very foundation of the system we are about to submit.

A tangent is a line touching a circle at right angles to the radius as shown at Fig. 14, and is readily constructed and as easily understood.

To construct Fig. 12, from center O with the radius OA, describe a quarter circle, APC; draw tangents AB and CB, join AC; through the point B draw a straight line parallel to AC; with center B, with radius BA, describe the arcs AD and CE; at the point E erect the perpendicular EF at right angles to DE to any desired height (in laying out a handrail this height will be the same as the height of the number of risers contained in the wreath); let F be the given height (this being one pitch); join FD, extend OB to G; from G draw GII at right angles to FD; make GH equal to

16 COMMON-SENSE HANDRAILING

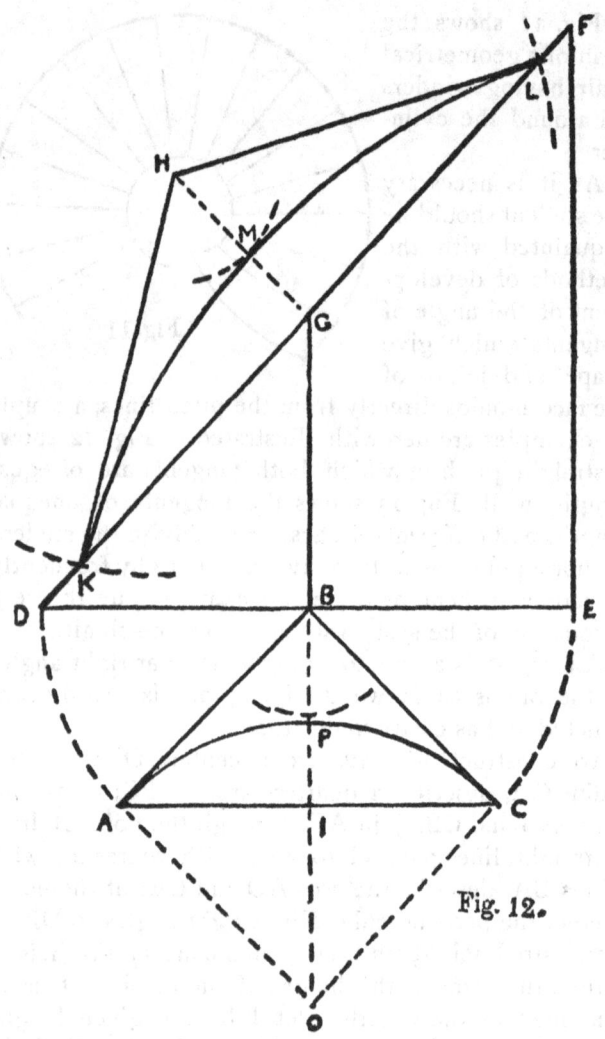

Fig. 12.

BI. With the center H and radius DG describe arcs, cutting DF at K and L; draw HL and HK, which are

FIRST METHOD

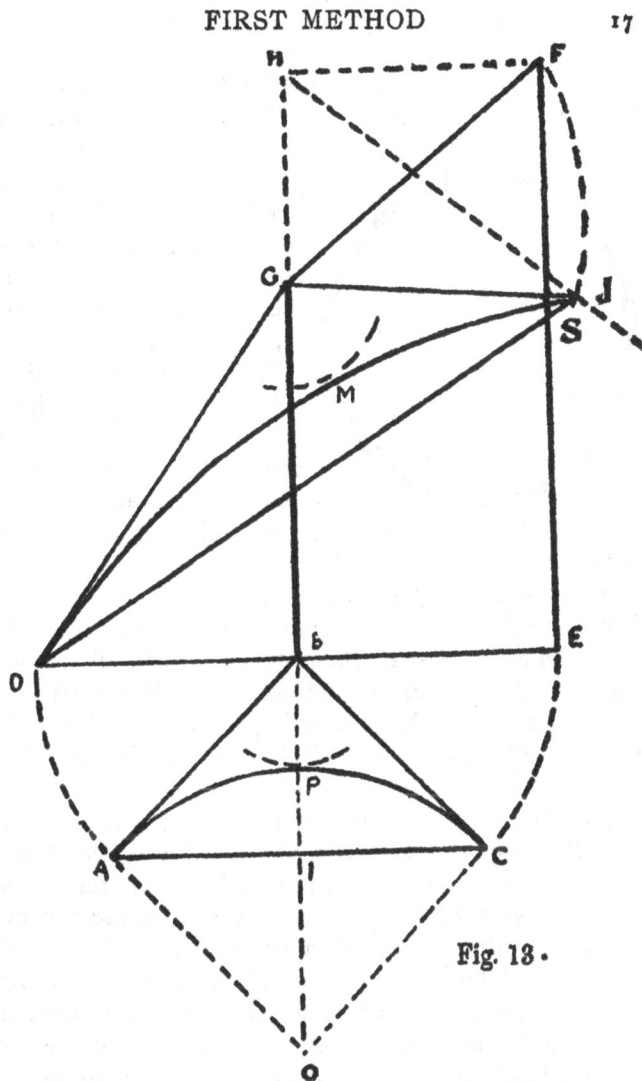

Fig. 13.

tangents on the pitch, and which, when placed in position, would stand plumb over ABC.

To construct Fig. 13, proceed in the same manner as in Fig. 12, until the height is located. It will be noticed that in this example BG is lifted higher, making the pitch-lines and tangents FG and DG of unequal lengths. To obtain the angle continue BG to H, making BH equal to EF; from H draw the line HJ to any distance at right angles to DG. With the center G and radius GF describe an arc cutting the line HJ at S; join SG and SD and the angle is completed.

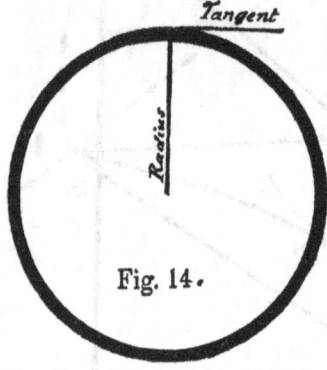

Fig. 14.

An easy way to prove the correctness of these problems is to draw them on common thick paper or cardboard on a larger scale than shown in these diagrams; then take a knife and cut out the angle DEF, place it perpendicularly over ABC, bringing D over A and E over C; then cut out the angle HKL, and if drawn correctly it will lie on the pitch-lines and fit the sides exactly.

To draw the curve line in the most practical way take B as a center, and with radius BP describe an arc touching the curve APC in the angle ABC; from H as a center, with the same radius, describe an arc cutting HG at M; then take a thin flexible strip of wood of an even thickness, bend it until it touches the points KLM; mark around it with a pencil, and the curve is completed, and near enough to absolute accuracy for all practical purposes. The curve so obtained in its perfection should be a portion of an ellipse, which it will be if correctly drawn.

FIRST METHOD

Let us now go back to Fig. 1 and describe the method for obtaining the face moulds and bevels of turnout and wreath pieces for that style of a stair.

To build these stairs correctly and with an easy, graceful rail, two or three things must be carefully observed in taking dimensions and laying down the plan. Measure the height from top of first, to top of second floor; set the rod you measure with plumb at the trimmer where the stairs land, and be sure that the lower end is level from where the stairs start. Measure the width of opening from studding to face of trimmer, also the depth of joist, that the cylinder may curve round and meet the face board level; plumb down from the header at landing, and measure back the amount of run where the stairs start; divide the height into the necessary number of risers, space off the run, making one less than in dividing the height, and also make allowance for the cylinder, landing and swell of the turnout steps. Where it is practicable make the rise seven inches or as near to it as possible, and make the tread, or step, ten inches or as near as can be, as this combination makes a very easy stair for dwellings, but of course the height of riser and width of tread will be dependent to a great extent on the surrounding conditions.

In laying out the steps for the turnout observe the same rule that applies to all winding stairs, that is, to make them as near the width of the straight treads as possible on the walking line. Locate the landing riser exactly half a step from the center of rail on landing, as shown at Fig. 15. This will bring the rail the same height on landing as it is in the middle of the step. Any departure from this rule will either change the height or will make it necessary to spring the wreath or slab off the shank, a very clumsy experiment.

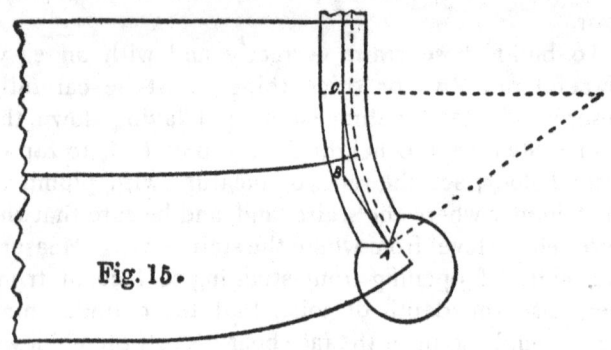

Fig. 15.

Fig. 15 shows the plan of turnout steps, with rail mitering into cap. The dotted curved line shows face of string. The black line shows center of rail with tangent ABC at right angles to dotted radius.

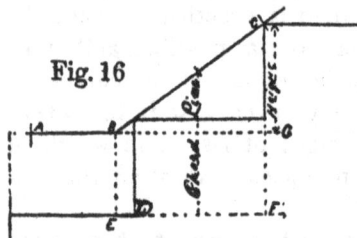

Fig. 16

Fig. 16 shows the tangents in position on the pitch. To construct Fig. 16, take the pitch-board and mark out the steps as shown. Beginning at third riser and coming down, draw pitch-line CB from second riser D; mark distance DE, which gives angle B; draw level tangent BA, agreeing with BA, Fig. 15; continue line of first step with dotted line to F, draw FC; continue the line AB to G. The distance from G to C is the required height, and EB gives the height to which the rail is lifted at the newel.

FIRST METHOD

To construct Fig. 17, draw tangents ABC, and curve line exactly like Fig. 15. In practice this figure can be drawn on Fig. 15, and to avoid confusion of lines it is transferred. Continue AB to D; draw DC at right angles to DB; set up the height, GC, Fig. 16. Connect ED at right angles to ED, draw DF; with D as center, describe an arc from B to G and from A to F, then connect EG; draw dotted ordinates AA to mark center of curve and chord line.

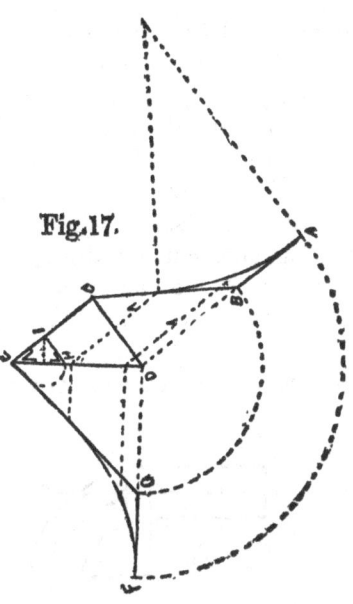

Fig. 17.

The spring bevel for squaring the wreath at lower end is found at angle E. To obtain the bevel for upper joints take a center anywhere on line ED, describe an arc touching EG and cutting ED at H. Draw line from center of arc at right angles to ED, cutting EC at I. Connect HI, and the angle at H is the upper spring bevel.

The development of pitch-line for wreath is shown at Fig. 18. First make a plan of the cylinder; draw center line of rail with tan-

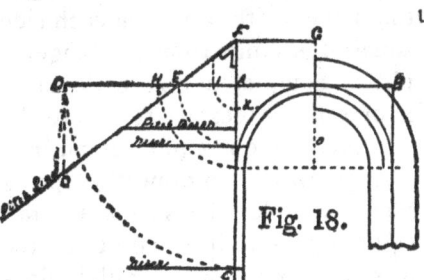

Fig. 18.

gents CAB (the distance from face of string to center of rail varies according to size of baluster), locate the risers, putting last one-half step from angle A, locate the joint of rail at riser C. With A for a center describe the arc ACD, extend AB to D, swing last riser around to E, chord line H, and X to I. Place pitchboard with riser touching AC, and hypothenuse or raking side cutting through E; draw pitch-line and continue AC to meet it at F. AF is the height of rail above the floor; draw FG at right angles to FC; continue radius O through to G, square up from I to pitch-line and from H and D down.

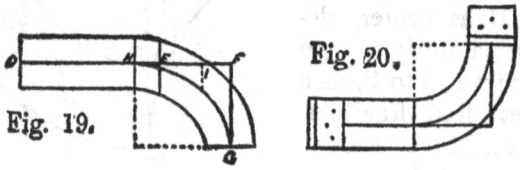

Fig. 19. Fig. 20.

To construct Fig. 19 (this figure can be drawn on Fig. 18, but is transferred for the same reason as Fig. 17), draw a line agreeing exactly with pitch-line DF, with points HEI marked; make FG at right angles to DF, and equal to FG, Fig. 18. Draw the line I equal to X, Fig. 18, bend in a thin strip of wood and draw curve GIH. Set off half the width of rail on each side of this curve line, square the joints from the tangents KFG, and the mould is completed. A little more than the finished size of rail is necessary to square the wreath, but not often more than one-eighth of an inch on each side. The surest way is to draw the spring bevel on a board, place a templet the size of the rail on a bevel line at right angles to it, square from the edge of board across corners, draw parallel lines

FIRST METHOD

enclosing the templet, and it will be seen at once how wide the mould should be and what thickness of plank is required. This method is seen in application of bevels at Figs. 21 and 22.

Fig. 20 shows the bevel portion of wreath; a better appearance is given to the wreath by using plank half an inch thicker than the rail, and casing it up from center joint as shown by the sections on end of mould.

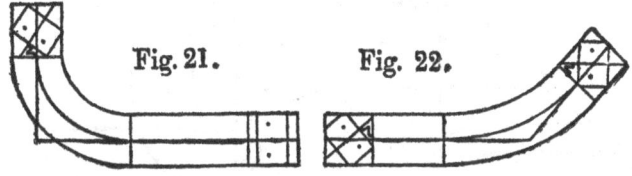

Fig. 21. Fig. 22.

Figs. 21 and 22 show the application of the spring bevels and templets for squaring the wreaths.

The bevel for Fig. 21 is found at F, and is simply the pitch of the stairs. The bevels for the turnout wreath are both applied in the same manner, from the inside, or the face cutting through the center, as shown by the sections.

The following illustrations exhibit a method of obtaining the face moulds for the flight of stairs shown at Fig. 2, which is a flight the most common in use.

To obtain dimensions make the plan, etc., and follow the instructions given for Fig. 1. Where the risers are located half a step from center of rail, as explained previously, the same method will apply to this flight, and the bevel will be the pitch of the stairs at the center point, and the section will be square with the face of stuff at the straight end of wreath.

Suppose Fig. 23 to be the ground plan of cylinder, with risers placed in a position that insures an easy, graceful rail, and also adds to the run by curving the landing and starting risers back to the platform.

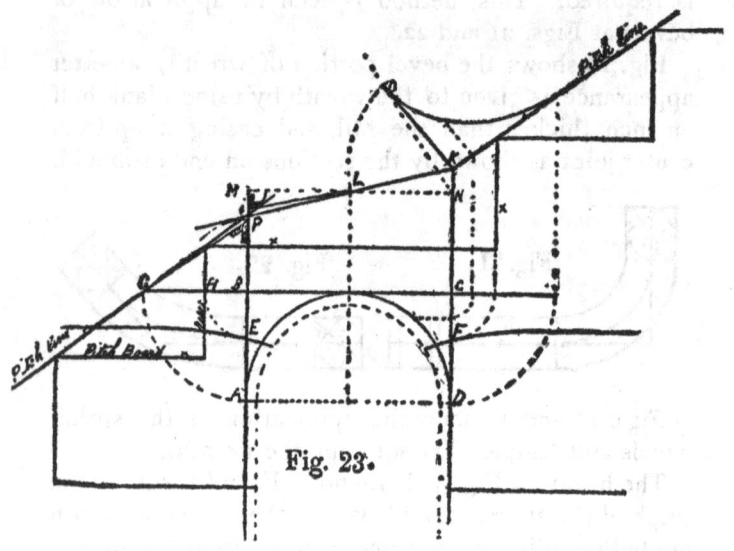

Fig. 23.

To construct Fig. 23, draw the center line of rail and tangents ABCD; from B and C as centers, swing around A and D, cutting B and C extended; swing around the risers E and F in like manner; place pitch-board with riser parallel to AB and touching H, and the raking side, cutting through G. Mark out the treads and risers as shown at XXXX. Draw pitch-lines, as shown, cutting AB and DC extended up; join IK, and the pitch-line is complete. To obtain the angle of tangents at K draw dotted line from center of cylinder cutting IK at L. Draw MN through L parallel with BC. At right angles to upper pitch-line draw dotted line NO. From center K swing around KL to O; connect KO, and the angle is complete.

FIRST METHOD

To obtain the spring bevels — from center B describe an arc, touching the pitch-line KI extended, and cutting BI at P; connect PG, and the bevel for center joint is found at P.

To obtain bevels for joints connecting with straight rail, take M as a center and describe an arc touching lower pitch extended connect with L, and the bevel is found.

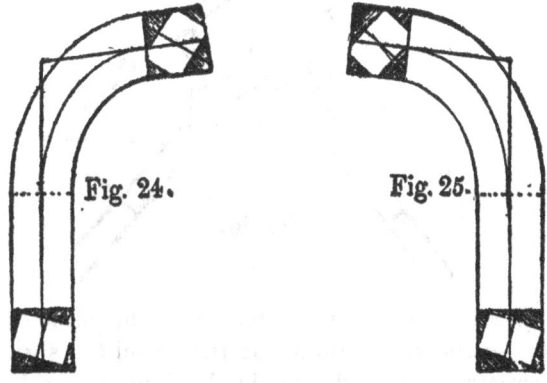

Figs. 24 and 25 show the sections and application of bevels on rails. Fig. 24 is the lower and Fig. 25 the upper wreath; the same face would serve for both, as the upper and lower pitches are the same.

Let us now examine Figs. 3 and 4, and endeavor to form rails to suit them. As before stated, these two examples represent on the ground plans obtuse and acute angles at the return landings; and in the formation of rails to meet the requirements for these stairs, the student will have covered the ground for the formation of rails for nearly every kind of rail required for a platform stair. In locating these risers at the landings be sure to place them, if possible, exactly half a step each way from angle B, Fig. 26. This will insure an easy rail.

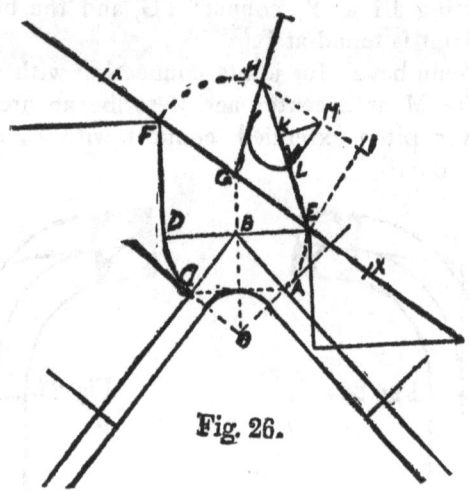

Fig. 26.

Fig. 26 shows the development of the angle of tangents for the face mould and the bevel for springing the wreath. Draw the angle ABC on center line of rail as shown; draw dotted line from center O to B; draw DE at right angles to OB; from center B swing around A and C to D and E; set up one riser from D to F, and one down from E; mark one step above and below the pitch-board; draw pitch-line XX.

Connect CA, and continue OB to G; with B as a center, describe an arc touching CA; from G as a center with the same radius, describe an arc; from E draw line touching this arc; from G again swing around GF to H; connect GH, and the angle of tangent is complete. The amount of straight wood on wreath is shown from E and H to the joints XX.

To obtain the bevel it is first necessary to find the

FIRST METHOD

point I; from H with FD for radius, swing an arc and intersect with another from E; having CA for radius, connect EI and HI; take a center, K, anywhere on the line EH, draw an arc touching GH and cutting EH at L; square down from K, cutting JH at M; connect LM, and the bevel is found for both joints of the wreath, the pitch being one straight line. Fig. 27 shows application of bevel to wreath.

Figs. 28 and 29 are simply a repetition of Figs. 26 and 27

Fig. 27.

excepting that the ground plan forms an obtuse angle.

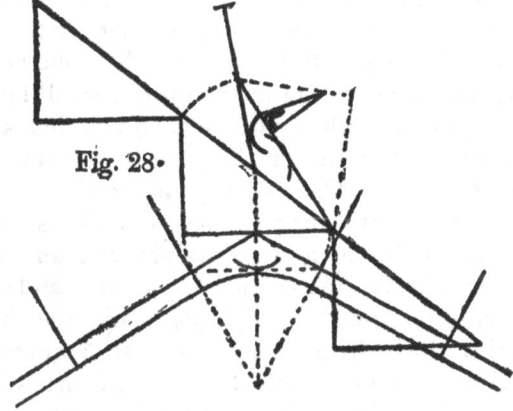

Fig. 28.

Fig. 30 shows the manner of sliding the mould on the wreath to mark it for blocking. We may state here that there have been worked some hundreds of rails during the past thirty years by this method, and we have come to the conclusion that the easiest and quickest way to block out a rail is to use just such

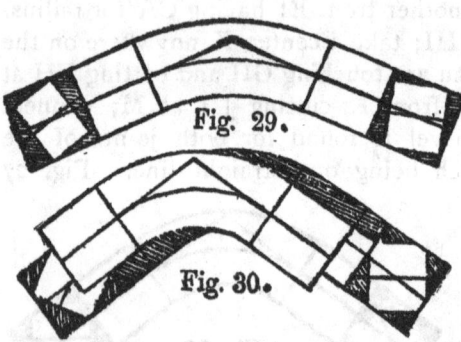

moulds as are shown in the drawing, viz., of a parallel width, and just sufficiently large to square the rail properly for moulding. When the wreath is sawed out, the face of stuff carefully planed true, the tangents marked and the joints made perfectly square with the face and with the tangents, then square the tangent across the joints, mark the center and draw the bevel across, as shown in Fig. 27, mark the section of rail at right angles to bevel. The best method of doing this is to use a thin templet with a small hole in the center, through which put a scratch awl, then swing the templet until exactly at right angles with bevel and mark all round it. The section being marked, square in from the joints on all sides to make sure the wreath will bolt on to the straight rail and form a clean line. To mark the curve line slide the mould up, as shown in Fig. 30, mark the inside edge (this line will not be quite as accurate as one made from an elliptical mould on the sliding principle with wide ends, but it is near enough for all practical purposes) by roughing out the inside first and occasionally planing through the wreath. Looking in the direction of a plumb-line, it will be seen at once when to take off the superfluous wood, and with a little care the inside will soon show a clean, true

FIRST METHOD

surface. As soon as this is done gauge the wreath to a width, then bend in a thin strip. Connecting the straight lines squared in from each end, mark around the outside in the same manner, mark off the top, gauge to a thickness, and the wreath is squared. The plumb-line can be marked on the inside of the wreath, and wil give the line of sight by taking the bevel from the angle EHJ, Fig. 26.

Now we will describe the method for constructing the face moulds of a handrailing for a stair suitable for the plan shown in Fig. 5. In this example two problems are used to obtain the development of tangents, bends and twists of the rail.

Let Fig. 31 represent the ground plan of cylinder with the risers marked in position, also the elevation and "pitch" inclination of center line.

It will be noticed that the pitch-line is perfectly straight. This is caused by the risers being placed so as to bring them exactly the width of a step from each other on the tangent line, as shown in the plan and elevation, Fig. 31. This is a point the student should always bear in mind; locate the risers this way whenever practicable, and you are sure to have a good-looking, easy rail.

To construct Fig. 31, the plan of cylinder being made with risers and center line of rail drawn, swing out A and D to meet BC extended, also the risers E and F; place the pitch-board at the point where E cuts the line BC, keeping the risers parallel with BA, and the raking side cutting through the point where A swings around to BC. Mark the step and riser and continue the elevation as shown; draw the pitch-line, draw GH, continue DC to L, draw MN at right angles to GH and extend to chord line at S: from N at right angles to pitch-line draw a line indefinitely; with L for center and LH for radius, describe arc cutting

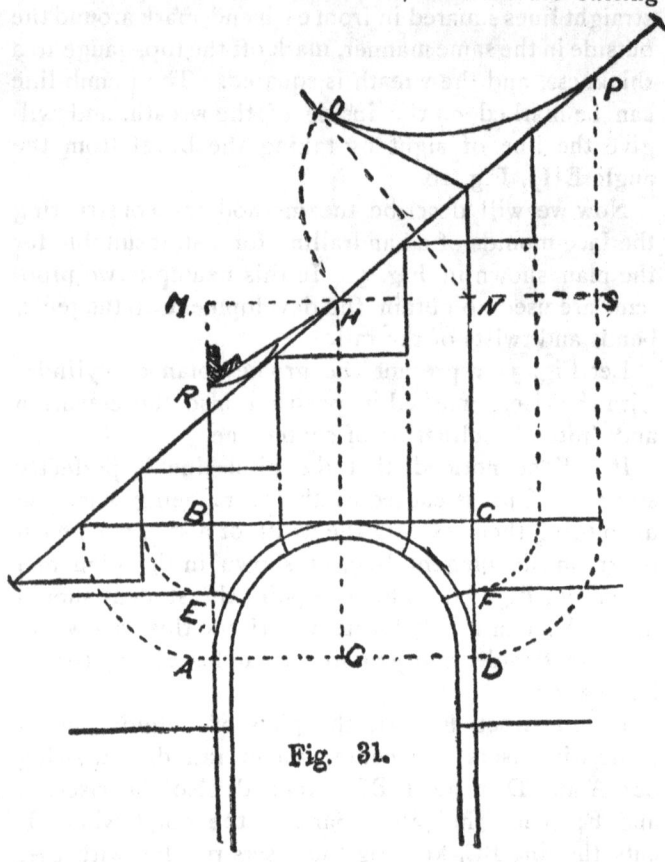

Fig. 31.

line drawn from N at O; connect LO, and the triangle OLP gives the tangents for the face mould.

The pitch-line being straight, the tangents are all of equal lengths, so it will be seen at once that the face mould obtained will be the same for both upper and

FIRST METHOD

lower wreaths, and the bevel for both ends is found at R, as shown.

Fig. 32 is similar to Fig. 12, excepting the development of tangents, which it will be observed is obtained somewhat differently.

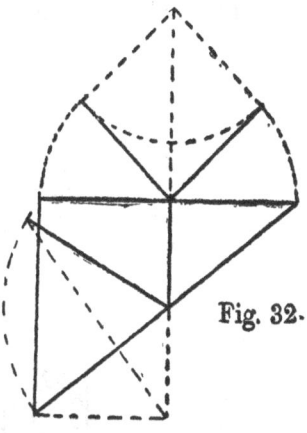

Fig. 33 describes how the development can be obtained by the method shown in Fig. 12. While this method is perfectly correct in all cases where the tangents are of equal lengths, still it is better to use the methods shown in Figs. 32 and 34, as they will be more correct whenever a change occurs in the pitch. Fig. 34 is a facsimile

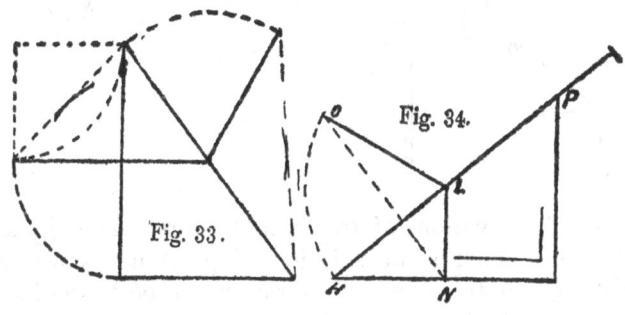

of the "development" in Fig. 31, and is drawn in order to make the student more familiar with this important problem.

32   COMMON-SENSE HANDRAILING

In order to produce the face moulds, bevels, etc., for the flight of stairs exhibited in the plan, Fig. 6, we must proceed as follows. Fig. 35 shows the ground

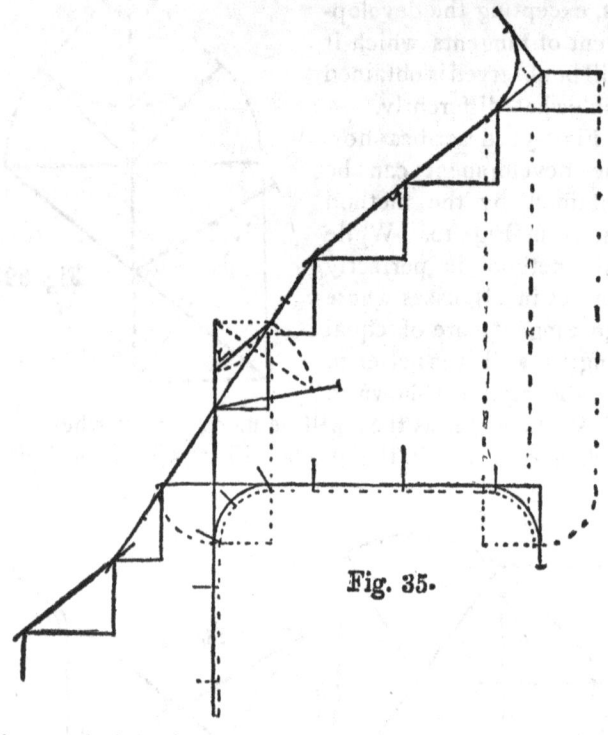

Fig. 35.

plan and elevations of treads and risers. The explanations given for the solution of previous examples will apply to this one if the figure be properly studied, as the method of proceeding to lay down the rail is exactly the same.

In the elevation it will be seen that one pitch is employed for the wreath and the connections made with the pitch of the flyers by a ramp above and below.

FIRST METHOD

One pattern answers for both ramps, as the pitch over the flyers is the same in both cases.

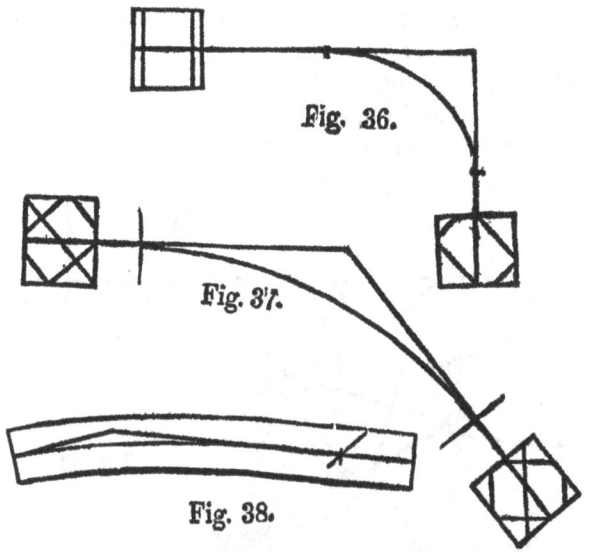

Fig. 36.

Fig. 37.

Fig. 38.

By carefully studying Fig. 12, the landing wreath, Fig. 36, will be easily understood. Care must be taken to locate the last riser as near half a step from the level tangent as possible.

Fig. 37 shows tangents, center line of rail and the application of the bevels for the wreath.

Fig. 38 shows pattern for lower ramp, and is simply reversed for the upper.

These examples are simple and ought to be readily digested by any workman who has ever had the least experience in stair-building. The young student who has never helped to build a stair or erect a handrail should master these simple problems (on paper) and the first opportunity that offers to see a flight of stairs and handrail set up he should embrace it, and the whole mystery of handrailing will disappear at once.

Figs. 39, 40, 41 and 42 represent the method of drawing moulds for flights of stairs similar to Fig. 7. Starting with winders in a quarter circle, Fig. 39 shows the ground plan of risers, also tangents around center line of rail, and their development. To con-

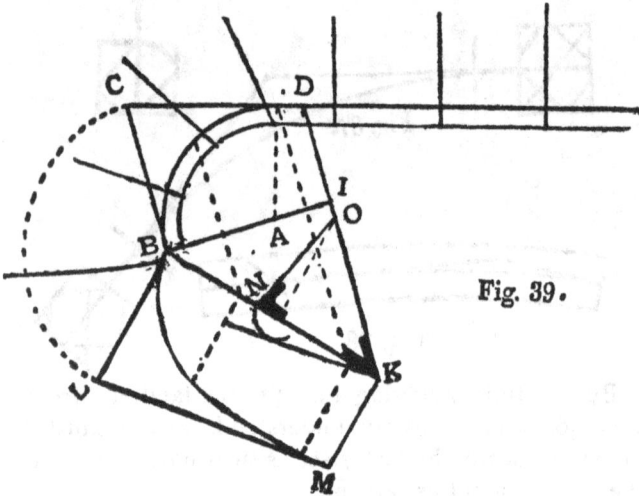

Fig. 39.

struct Fig. 39 draw radius from center A to joint of rail at newel B. At right angles to AB draw BC. This will give the angle of tangents on ground plan. In wreaths of this shape the tangent BC is always level so as to give a plumb joint at the newel.

Before proceeding further with Fig. 39 the height must be obtained by drawing Fig. 40; this is done in the manner explained in previous examples, viz., by setting up each riser and obtaining the width of the treads from the tangents BCD, where the risers cut through. To develop the mould, Fig. 39, extend BA

FIRST METHOD

to I at right angles to BI, and from joint D draw DK. Make IK equal to Fig. 40. Connect BK at right angles to KB and draw BL and KM. BL should equal BC, as shown by the arc, and KM should equal ID. Connect LM and the angle is formed for the mould. The dotted ordinates give the springing line and a point through which to bend the strips to obtain the curve. The bevel for lower joint is found at K. To obtain the bevel for the upper joint draw a line from K parallel to ML. At any point on line KB describe an arc touching the line drawn from K. Draw a line from center of arc at right angles to KB, and cutting KD at O. Connect ON and the bevel is found at N for the upper joint.

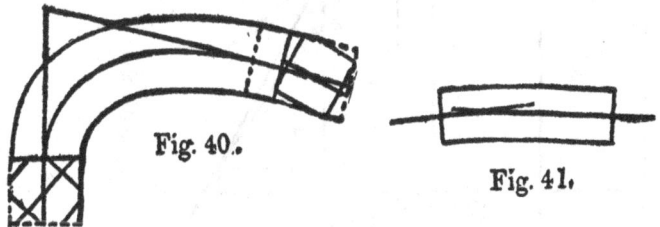

Fig. 40.

Fig. 41.

Fig. 40 shows the application of bevels, the upper bevel cuts shown through from the outside in all cases where the angle BCD, Fig. 39, forms an acute angle; when the angle is obtuse the bevel is applied from the inside, the lower bevel in all cases remains the same in application.

Fig. 41 shows the ramp.

There seems to be no difficulty presented in these problems that cannot be readily overcome if the student but applies himself diligently. We would suggest that each one of these figures be drawn and redrawn, until the student has become so familiar with each one of them that he can draw them from memory alone. Such practice will not make very serious inroads on his time, and what investments in time are made will, in the not very distant future, return big interest.

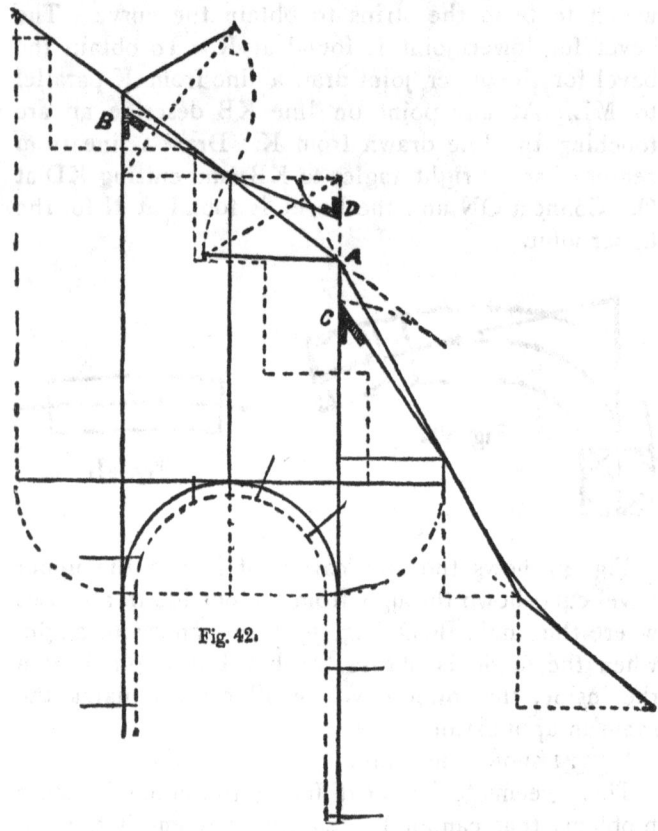

Fig. 42.

In Fig. 42 we represent a flight of stairs with four winders, quarter landing, and return flyers.

FIRST METHOD

This shows the ground plans of cylinder, with risers cutting the tangent around the center line of rail; also the elevation of risers, the pitch-lines, and the development of the tangents.

It will be noticed that the upper pitch-line is the same as that of the return flyers, running down until it meets the lower pitch at A. The rail is lifted higher than usual at this point, but this is a defect which will not detract from its appearance, and makes a much better wreath than is drawn with a ramp.

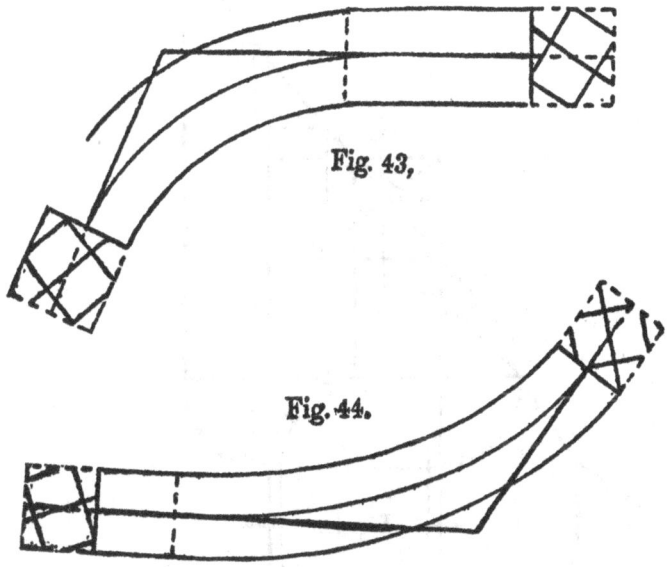

Fig. 43,

Fig. 44.

Fig. 43 shows the face mould for the lower wreath, and Fig. 44 for the upper, also sections of rail and applications of the bevels.

The bevel for upper wreath is found at B, Fig. 42, as shown in the diagram, and is the same for both joints.

38 COMMON-SENSE HANDRAILING

The bevels for lower wreath are found at C for upper joint, and at D, Fig. 42, for the lower joint.

The plan of this stair for which the rail is intended

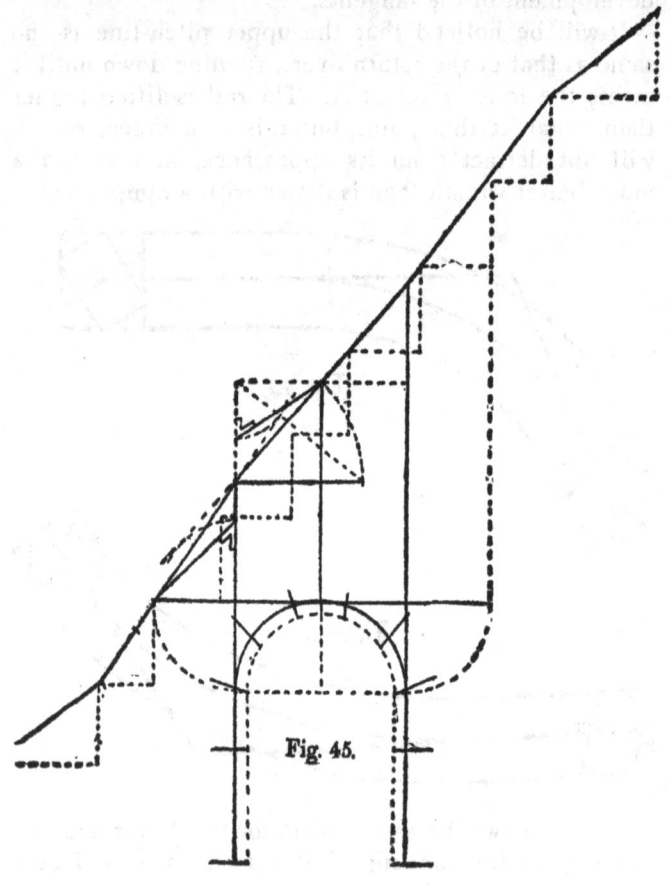

Fig. 45.

FIRST METHOD

is shown at Fig. 9, which, upon examination, will be found to be of a type often found in our old colonial buildings.

We will now deal with a flight of stairs having seven winders forming a half circle, with flyers above and below.

Fig. 45 shows the ground plan and elevation of risers, the pitch-lines and development of tangents for face mould; the upper and lower pitch being the same, only one face mould is required. The mould is simply reversed in application, as shown in Figs. 46 and 47.

Further explanations of these figures

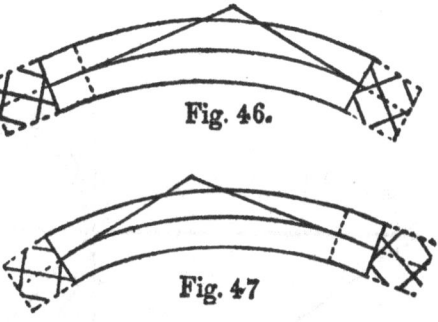

appear unnecessary, as the lines and applications are similar to those applied in previous examples, so that a reference to previous illustrations and the descriptions and explanations attached will give a clear insight into the method of lining out the present examples.

The next examples show a method of laying down the rail for a flight of circular stairs. Fig. 48 shows the plan with cylinder and risers cutting around the center of rail and tangents. The joints are located at A, B, C and D, making four pieces; the two wreaths from A to C are alike, and only one mould is required.

The rail is in one pitch from E to A. Fig. 49 shows the elevation of steps and risers for first wreath,

40 COMMON-SENSE HANDRAILING

and Fig. 50 the landing wreath; these are drawn in the manner as shown in previous examples, making

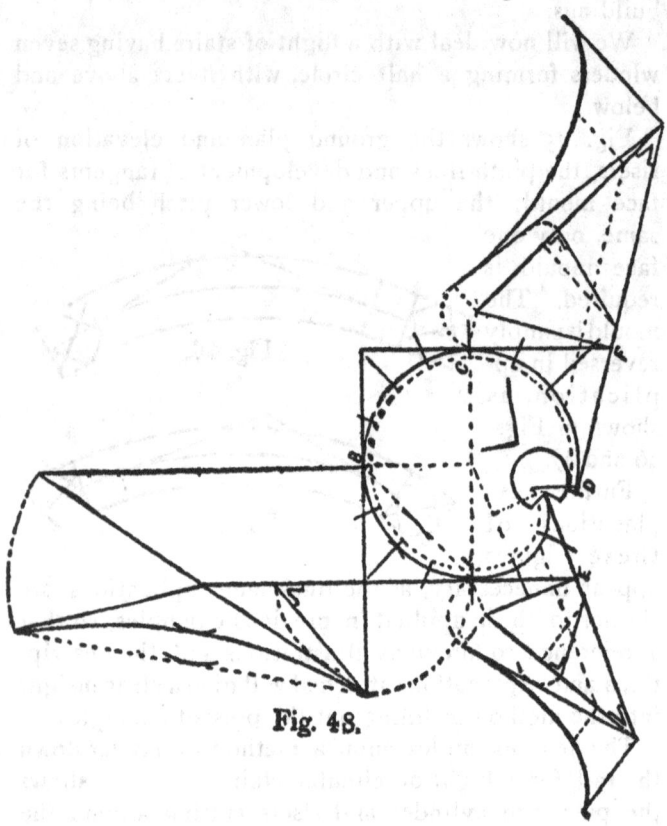

Fig. 48.

the width of treads to correspond with the points where the risers cut the tangents in Fig. 48. These elevations also give the exact height as shown; the first wreath in Fig. 49 is lifted for the newel, and the landing wreath in Fig. 48 runs half a rise above the floor.

The wreaths AB and BC simply rise from risers,

FIRST METHOD

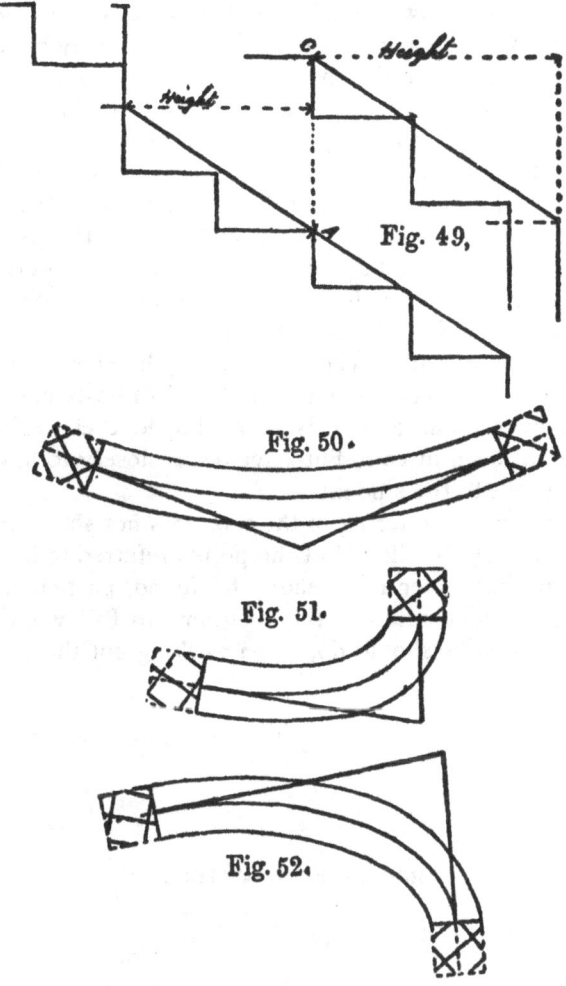

and are drawn as shown in Fig. 48. Figs. 50, 51 and 52 show the face moulds and application of the bevels, in the same manner as explained in previous illustrations.

We have now completed the treatise as first intended, and have shown how handrails may be laid out and made for eleven different styles of stairs, and from the rules given the student should be able to lay out a rail for almost any kind of rail he may be called upon to construct. The principles involved in this method of handrailing are well described in the earlier part of this work, and we would advise the young reader to again and again go over them, and produce the lines on a good-sized drawing board until he becomes familiar with the methods.

We have already given diagrams of the eleven kinds of stairs, and have now shown how handrails may be built over them, and it is to be hoped these efforts have not been in vain, but have been closely followed with profit to the student.

Some of the lettering in the cuts does not show up as well as would be liked, but the points referred to in the descriptions given may readily be found, particularly, if the student enlarges the diagrams to full working size—which he should do, when working out the problems.

END OF FIRST METHOD

SECOND METHOD

The method of laying out handrails shown in this section differs somewhat from the method shown in the previous section, and has some advantages the former does not possess. As this little book, however, is intended for instruction and not to advance the interest of any particular method, the editor and compiler has thought fit to present to the reader several methods—all of them of the simplest sort—in order that he may find something he can utilize in each and all of them.

Before building a handrail it is quite necessary to have the stairs, and as the "handrailer" is supposed to know how to construct the body of the stairs, we shall content ourselves with making a few remarks concerning the height of riser and width of tread.

After determining the height of the riser from the "story rod," the right proportion of tread must be found. Sometimes steps are arranged so that it is easier for a man to go up "two at a time" than to walk up in the proper manner. The reason is both tread and riser are made small. When a riser is reduced the tread must be increased; and the contrary, when the riser is increased, the tread must be reduced in width. Joiners do not often break this rule, but masons very often do, notably in steps leading to and from railway stations. A simple rule may be given for finding a suitable proportion.

Take any suitable step as a standard step, that is to

say, if you know of a staircase which is comfortable and easy to walk up, take it as a standard to gauge others by. Suppose you have a riser given, and require the width of a suitable tread, make use of the following proportion:

As the given riser : standard riser :: standard tread : required tread :

If the tread is given and the riser required, then:

As the given tread : standard tread :: standard riser : required riser.

To work out an example, suppose 10-inch tread and 7-inch riser be taken as a suitable step. Let 6 inches be the given riser; then by substituting the value of treads and risers for the names we have, as 6" : 7" :: 10" : the required tread. This gives $\frac{70}{6}$ or $11\frac{2}{3}"$ for the size of a tread. Nicholson gives as a standard a tread of 12" to a riser of $5\frac{1}{2}"$. Working out the example given by this proportion we get 11 instead of $11\frac{2}{3}"$; either of these sizes will be an agreeable step.

A rough and ready rule for the usual sizes of treads and risers is to make 2 risers and 1 tread equal to 24 inches.

Before going into the working part of stair-building it must be understood that great care ought to be taken in placing the staircase in any building, and, therefore, staircases ought to be described and accounted for justly, when the plans of a building are made, and for the want of this, sometimes unpardonable errors are made—such as having a little blind staircase in a large house, and on the other hand a large and spacious staircase in a small house. In placing staircases the utmost care ought to be taken, it being a difficulty to find a place convenient for them, that will not at the same time prejudice the rest of the

SECOND METHOD

building. Commonly the stairs are placed in an angle, wing, or middle of the front. In every staircase openings are required—first, the opening leading thereto; second, the window or windows that may give light to them; third, their landings. First, the opening leading to the staircase should be so placed that most of the building may be seen before coming to the stairs, and in such a manner that it may be easy for any person to find them. Second, the window must be placed in the middle of them, whereby the whole of the stairs may be lighted. Third, that the landing should be large and spacious for the convenient entering of the rooms—in a word, staircases should be spacious, light, and easy to ascend. The height of risers should be from 6 to 7 inches, the breadth of tread not less than 9 inches, and the length about 3 feet—the rule laid down for the height and breadth of steps. Workmen are, however, not to be so strictly tied to those rules, as shown above, as not to vary in the least from them. They must endeavor to make all the steps of the same staircase of an equal height and breadth. To do this they must first consider the height of the room, and also the width or compass they have to carry up their stairs. To find the height of each step they ought first to propose the height of each step, and by that proposed height divide the whole height of the room, which done, the quotient will show the number of steps. If there is a remainder, then take the quotient for the number of steps, and by the number divide the whole height of the room, and the quotient will be the exact height of each step.

Example: Suppose the height of the room is 9 feet 3 inches, and you propose your riser to be about 6 inches; bring the height of your room into inches and divide by 6 inches. You have 18 steps and 3 inches over, therefore, take 18 for the number of steps and by it divide 3 inches. The quotient will be $6\frac{3}{18}$, or $6\frac{1}{6}$, which must be the exact height of each riser. You find the breadth of steps in a like manner.

Having determined the height and breadth of your steps you then make a pitch-board which is a triangle of unequal sides, one being equal to the breadth of step, the other equal to the height, thus giving the rake of stair.

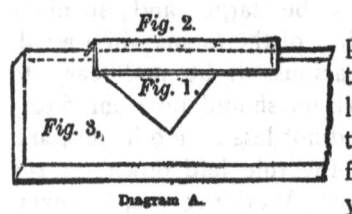

Diagram A.

Fig. 1 (diagram A) is the pitch-board. Fig. 2 is a templet about 18 inches long, $2\frac{1}{4}$ wide, which is used to form a stop or gauge for the pitch-board when you are setting up your steps. Fig. 3 shows the templet and pitch-board applied to plank intended for the wall string.

In the formation of winding stairs much care must be exercised in laying them out. The following diagrams show a stair with six flyers and six winders, with instructions to lay out and set up the strings.

It must always be understood that you must lay down a plan of your winders, the full size the pitch-board will give the flyers. Diagram B is plan of winders. Fig. 1 is the first wall string. Set up the first three steps with your pitch-board, then set up one riser; take the width of first winder on plan and mark it on the string square with the riser; then set up another riser and take the width of your other winder up to the angle, and mark that the same way. This angle winder is called the kite winder. You must then allow the string about ¾ of an inch longer for a tongue to go into the cross-string; then cut the string off at right angles with the step and allow about 6 inches from the step upward to form the top easing to carry

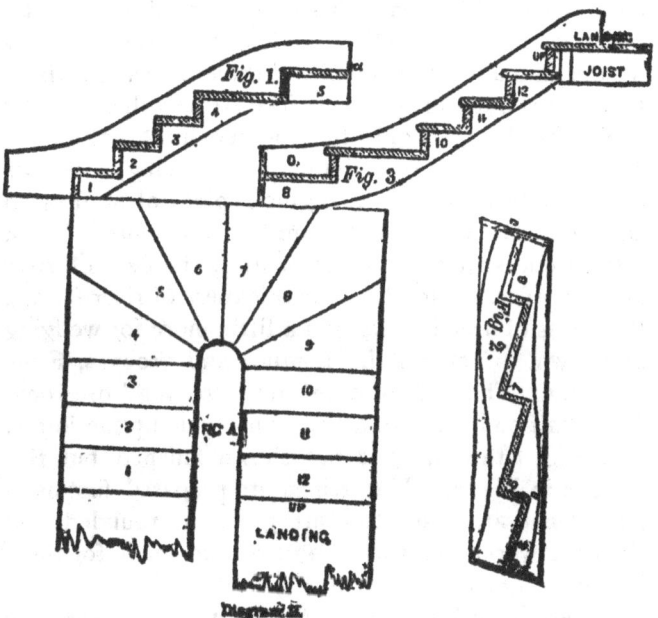

out the winder. You will see that a piece must be glued on the under side of string. I have shown easing at top, and also shown ramp at bottom to receive base, etc.

Fig. 2 is the cross-string. Always glue up cross-string for stairs of this description 14 inches wide, and then make a line, AB; from that line square off the end of string. There is no particular rake for the line, it being marked at pleasure. After squaring the end from the line you must set in the thickness of the risers and treads as shown in Fig 1, then set in the other halt of kite winder, then set up a riser square with the winder, set up the other winders and the half winder square with the end; then allow for tongue, etc. There will be enough stuff to form all easements.

Fig. 3 shows the other wall string having half a winder and a whole winder and three flyers, and risers that carry up to the landing. The string will be set out similar to the first, only the up-risers must not be forgotten. Groove the winder end of the string to receive tongue of cross-string; also glue a piece of stuff on to carry out the winder and form the easements. When setting out strings the pitch-board is the face of riser and top of tread, so allow for thickness of riser *in*, and thickness of tread *down*, and a little more for wedging as shown. C shows the tongues and grooves, F the wedging. The general depth for "housing" or grooving is half an inch. In all cases use glue up the joints.

In Fig. 1 the string is not shown finished, but this is done in Fig. 3. The strings are prepared first as in Fig. 1, and after the steps are glued up, rounded, and the cove worked on them, mortises are made for them in the string as shown at Fig. 3.

"There is nothing new under the sun," said the wise king of old—not even in handrailing, though much has been written on the subject since the celebrated Peter Nicholson pointed out the true theory of laying out this sort of work; yet, notwithstanding all the knowledge acquired since Peter Nicholson wrote, the art of handrailing has been a sealed book to nine-tenths of otherwise good joiners, and to-day it is often difficult to find a man who is not a professional hand-railer, who is willing to undertake the building of a rail over a circular staircase. This distrust, or per-

SECOND METHOD

haps lack of knowledge, exists even among the best and most competent workmen, and is a great retarding factor which ought not to exist.

In order to assist those who are desirous of studying this beautiful art I will submit such problems and their solutions as I may think will be of the greatest service, and which I may be able to select from the material at my disposal.

The proper construction of stairs is an all-important part of house-building to both architects and owners, as their daily and hourly use affords comfort and ease, or tires and distresses, as the case may be, according to the accuracy with which the true principle of stair-building is observed. Strength and solidity are also important factors, especially in stairs subjected to severe usage in the passage of safes, heavy trunks and weighty packages, as in cases like that in France, where, by the falling of a stone staircase a short time ago, the lives of scores of unsuspecting people were destroyed and a large number maimed and crippled for life as a result of a defective construction and the overconfidence of a thousand human beings who risked their lives upon a thing of beauty which collapsed from overloading.

I think there are not more than three flights of stairs of similar construction to that just alluded to in the United States, the most notable and prominent of which is in the State House at Columbus, Ohio. Such stairs are supposed to be self-supporting, the wide ends of the steps being inserted in the walls of the stairway, and the cylinder ends of intermediate parts lapped and cut, locking the respective steps together, so as to provide a continuous strength to the entire construction.

50   COMMON-SENSE HANDRAILING

The following diagram C shows a very easy way of getting out stair-rails over cylinders, 1-1-1 being the face of cylinder, 2 the center of baluster from which the tangents must be derived, and is, in fact, one of the most important points in the ground plan, for the reason that the rail is supported all around the cylinder on the top and center of the balusters.

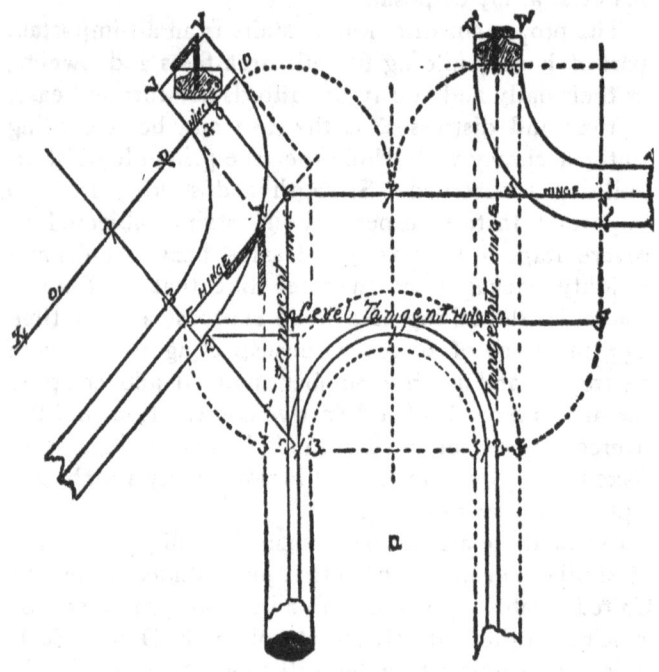

DIAGRAM C.   A RAIL AROUND A CYLINDER

Commence by making the tangent as shown; place half of the rail each side of 2-3-3; draw level tangent right and left indifferently; take the compass, stand

SECOND METHOD

on left, describe a curve from the spring line at 2, to cut level tangent at 5; do the same on opposite side.

The tangent being unfolded, set up the highest—say from 4 to 6 on the left; connect 6 and 5, which will give the tangent in the pitch; square out from 6 to 8, and stand compass in 6; open to 7, the center of cylinder; draw curve, cutting at 8; continue line 8 to 9, parallel with pitch 6-5; 8-9 then becomes the place of the elliptic curve covering the pitch. The breadth of rail in pitch is next to be found, as shown by the illustration, carried up from 3-8, cutting pitch-line.

Now to find the pin points to draw the elliptic curve, take the semi of the major axis (or half the length) from 9 to 10, place the compass in 12, and cut the major axis at XX,—XX being pin points for the outside curves.

Take the distance 9-11 in compass and stand in 13 and cut 0-0 on the major axis, which are the points to draw the inside curves by. The bevels are shown and how applied on the face of the moulds.

The cutting and standing up is explained as follows: Cut on line from 4 to 5 on either side, and from 5 to 13; from 13 to points marked V, V, V; from V to 6; from 6 to points marked V′,V′,V′,V′,V′,V′,V′; from V′ to 5; hinge on lines so marked and stand in place.

This is a very simple solution of what often confronts the young workman when building his first handrail. This, of course, is intended for a straight stair having a small well hole and landing on a level floor.

Problem.—To obtain a wreath for half-space landing: When the distance between the centers of rails is equal to, or more than, the width of a step, and the risers placed half a step from the center of rail at the crown of well, the wreaths for this class of staircases are of the simplest kind, being beveled at one end only, as the tangent line across the back of the

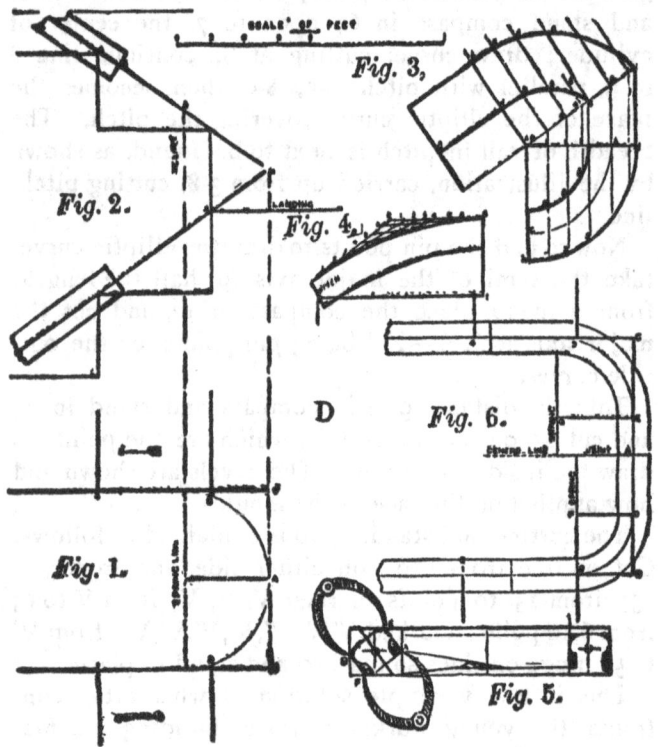

DIAGRAM D. HALF-SPACE RAILING

well is a straight level line. Fig. 1, diagram D, shows a large well with the risers 2 and 3 placed half a step from B and C; these four steps are shown in section at Fig. 2. Draw the center lines of rails 1,

SECOND METHOD

2, 3 and 4, and they will meet at A, which is over A, at Fig. 1; or the risers at Fig. 1 may be in any position, so that 2B added to 3C equals the width of a step; if the distance between the center of rails is equal to a tread the risers 2 and 3 would, of course, be at the spring line. At Fig. 6 is shown a wide well with the risers 2 and 3 half a step from B and C, with the radius of center line of rail equal to half a step as 2F, or more than half a step as GE. The wreaths may be jointed at A, the face moulds for these wreaths would be drawn as at Fig. 3, and the straight length AE or AD at Fig. 6 drawn square from E and F at Fig. 3, and worked to the bevel at Fig. 5 and out of the same thickness of plank. The center of the rail on the landing would be equal to half a riser higher than the center of the rail plumb with the risers. The wreath at the start of stairs from a landing is drawn as at Fig. 3, and then turned the other side up. Fig. 3: Mark the plan of wreath and the center line, BC; draw the lines AB and AC square to each other; draw AD with the pitch-board; draw ordinates 1, 2, 3 at any distance apart parallel to AC; draw the perpendicular lines, 1, 4; draw the lines 4, 5, 6 square to AD; apply the lengths, 1, 2, 3, to 4, 5, 6, and draw the face mould through the points 5, 6; make the shank, JG, about 9 inches long parallel to AD. Fig. 4: Draw HI equal in length to outside the plan of rail, HI, at Fig. 3; make ID equal to BD, at Fig. 3; make BD and HA each half the thickness of rail square to the pitch; draw AC with the pitch-board, and draw BC parallel to HI; divide AC and CB into the same number of equal parts, and join the points 1, 1 and 2, 2, etc., which will give the top curve, and gauge the thickness of rail from the top curve.

54 COMMON-SENSE HANDRAILING

To square the wreaths: First, cut them out square and joint the ends square. Second, center the joints, apply the bevel IFE at Fig. 3 (which is set to the top corner of the pitch-board), through the center of the top joint, and mark the square section of rail, as at Fig. 5. Third, apply the face mould on the top of the wreath and slide it up the shank until E, at Fig. 3, comes to E, at Fig. 5, and mark it; apply the mould on the under side until F, at Fig. 3, comes to F, at Fig. 5, and mark it. Fourth, set a pair of calipers to the width of the rail, and move one arm along the mark made by the face mould on the under side of wreath, and the other arm will mark, very nearly, what to take from the inside as at A, A, A, Fig. 5, and cut it out with a band saw; then gauge the outside from the inside with the calipers. Next work the top, gauge an equal portion of the top and bottom of the shank, as at Fig. 5, and lap Fig. 4, falling mould around the outside; place the shank in the bank, screw up to the right pitch and work the top level across in a direction toward the center of the well around the curve; apply a square to the top with the stock plumb on the inside. Gauge the bottom from the top. The top wreath is worked to the same hand and turned over before doweling or moulding. To joint the straight rails apply the length IG, at Fig. 3, from C and D at the spring line of well, at Fig. 2, to B and E.

Problem.—To obtain a wreath for a quarter-space landing: For small wells with the riser, A and B, *less* than half a step from the point C, where the space for the steps is confined for room.

Fig. 1, diagram E, shows the center line of rail and the plan of stairs.

Fig. 2: Draw the tread, DE, and riser, EA; make

SECOND METHOD

ACB equal to ACB, at Fig. 1; draw the riser, BF, the tread, FG, and the riser, GH; draw DA and CM; draw HF to M; draw MJ so that the part LJ will measure 5 inches; mark the joint, K, 2 inches from L; make JI equal to JK, from I to K draw the dotted lines square to IJ and JK, and from the point where they intersect

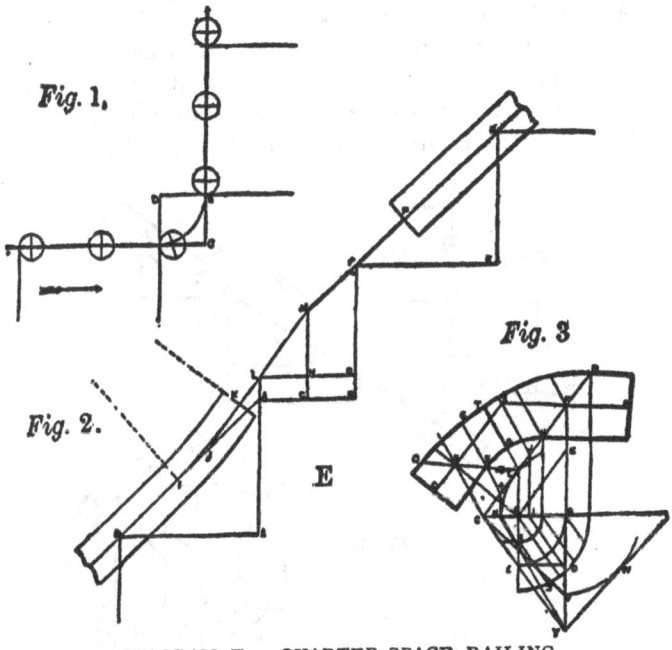

DIAGRAM E.   QUARTER-SPACE RAILING

describe the easing; joint the top rail square at D about 4 inches from F; draw LNO.

Fig. 3: Mark the quadrant and the center line, BC; draw the square, ABCD; set up BE and BF equal to NM and CF, Fig. 2; draw AE, and draw FG parallel

56 COMMON-SENSE HANDRAILING

to AE; draw AH square to AE from G. With the length, GC, cut the line at H, and draw GH.

Draw ordinates 1, 2, 3 parallel to GC; draw the perpendicular lines 1, 4, and the lines 4, 5, 6 parallel to G H; apply ID from J to K, and draw the tangents HK and KF; make HO and EP equal to LK and FP, Fig. 2; make the joints square to the tangents.

Apply the lengths, 1, 2, 3 to 4, 5, 6, and draw the face mould through the points 5, 6; draw the line 4HQ; make HQ equal H5, and from Q5U and R draw the shanks parallel to the tangents; continue GC and BD to meet at Y; from B describe the arcs LN; draw

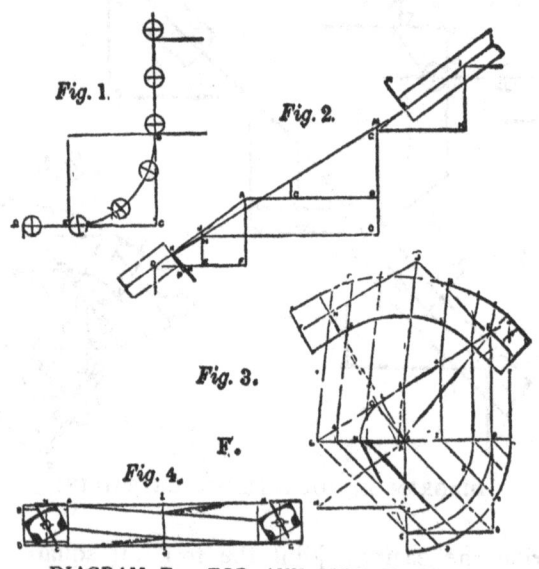

DIAGRAM F. FOR ANY SIZE CYLINDER

NY, and BNY is the bevel for the bottom end; make BM equal BF; draw MY; from B describe the arc, WV; draw VG, and BVG is the bevel for the top end.

SECOND METHOD

Cut the wreath out square to the plank and a little full in the narrow part. Apply the mould and square up the wreath, taking an equal portion off the top and bottom, both inside and outside the wreath at the line ST, Fig. 3.

Fig. 1, diagram F, shows a better plan of stairs for a large well than Fig. 4, as there are *two* balusters on the landing, the same distance to the centers as those on the steps, the wall bracket and nosing are much larger, and the steps are not so confined at the well end. By this method the easing on the straight rail for small wells, where the radius of center line of rail is less than half a step (as in the last diagram), is dispensed with.

Fig. 2: Draw DEF equal to DEA, Fig. 1; draw FA, a riser; draw ABC equal to ACB, Fig. 1; draw the riser, BG, the tread, GH, and the riser, HI; draw DA and GI. Joint the straight rails square at K and L about 4 inches from J and G; draw the falling line, KL; draw EN; draw NO parallel to AB; draw KQ square to AK, and equal to half the bevel line, NC, at Fig. 4.

Fig. 3: Mark the quadrant and the center line, BC; draw the square, ABCD; set up BE equal to OM, Fig. 2; make AG equal to AC; draw EG and GC; draw AH square to GE from G. With length, GC, cut the line at H, and draw GH; draw the ordinates 1, 2, 3 parallel to GC; draw the perpendicular lines, 1, 4; and draw the lines 4, 5, 6 parallel to GH.

Apply AD from I to J, and draw the tangents, JH and JE; mark A to P, a riser, and PM, a tread; draw MA; draw AO square to MA; draw OH; continue the lines HO and JI to meet at R, and draw RE; make HF and EK equal NK or ML, Fig. 2; draw the joint

F parallel to HR, and the joint K parallel to RE; apply the lengths 1, 2, 3 to 4, 5, 6, and draw the face mould through the points 5, 6; from A describe the arc, LN; draw NC, and ANC is the bevel for both ends.

Cut the wreath out square to the plank and a little full in the narrow part; cut the joints at first square to the plank and the length, PQ, Fig. 2, longer at *both ends* than the mould; apply the bevels and the mould, and work the inside and outside of the wreath as described previously. Now cut the bottom joint to the bevel NKQ, Fig. 2, applied to the *beveled sides* of the wreath, with the bevel stock held parallel to the tangents on the wreath and apply the same.

THIRD METHOD

This method is one much used by English and German handrailers in Europe, and as it is based on the system formulated by the late Robert Riddell, it is also practiced by many handrailers in America. The system has been very much improved and simplified by Mr. John Wilson, and with the exception of a few additions and corrections it is his version of the system that is herewith reproduced, and I am sure the student will find the matter as set forth in these pages clear and easy to understand, as everything of an abstruse character has been eliminated.

The upper portion of the fence formed on the outside of the stairs is the handrail, to assist in ascent and descent of the stairs, and also for protection. It is evident that the rail should follow the line of nosings and at a height of 2′ 9″ to the top side of the rail from the tread at the nosing, measured perpendicularly in line with the face of the riser.

In the construction of handrails the chief difficulty is in the wreaths, where the rail is of double curvature. Simple curves in either plan or elevation will cause no difficulty.

Fig. 1 shows the plan of a rail for a level landing stairs with the risers landing and starting in the springing, the radius of the center line of rail half the width of tread.

Having the plan and center line drawn, the wreath being in two pieces and one face mould answering for both pieces, to draw the face mould. First draw the joint line CD, Fig. 1, then draw the tangent lines, **AB**

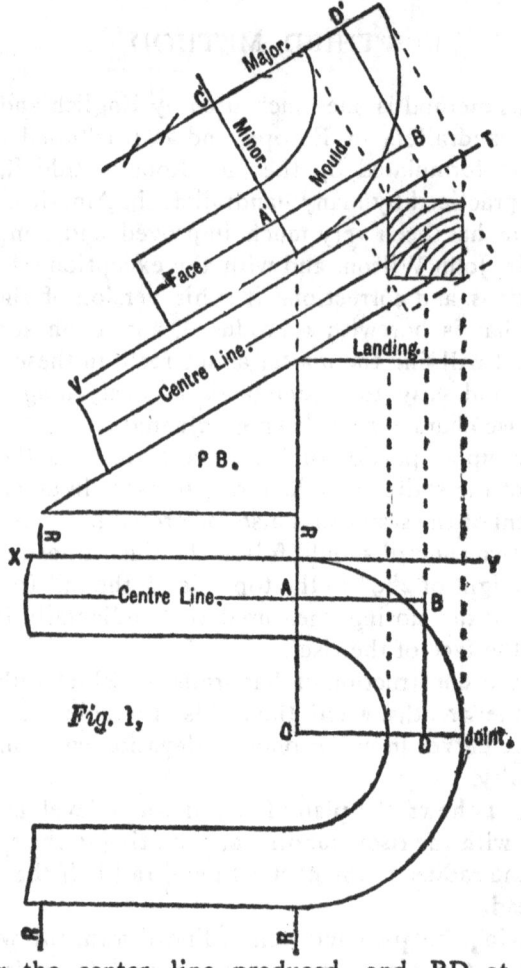

Fig. 1,

being the center line produced, and BD at right angles with the joint line. AB being equal to half the

THIRD METHOD

width of tread, the rail in coming up over this distance would rise the height of half a riser, causing the line BD to be horizontal; where one of the tangent lines is horizontal it is at once used as the directing ordinate. Then produce BD, and parallel with it draw lines from each side of the rail at D; and from the springing at A draw XY at right angles to BC, and place the pitchboard as shown against the riser line. Draw the under side of rail, set off half the depth of rail and draw the center line. Where it cuts the line from BD, will be the center of the section on the landing. Draw the section as shown and through the top corner draw VT. From where the lines projected from the plan cut this line. Draw lines at right angles. On each of these lines mark off the corresponding distances in plan, measuring from XY. Through the points draw A'B', C'D', and A'C', B'D'; C'D' is the major axis line, and A'C' the minor. The lines on each side of D' give the semi-major axis for the inside and outside curves for the mould, and on the minor the width of mould is the same as in the plan.

Draw the curves with trammel or string and pins. The shank may be made any convenient length.

Referring to the section at the line VT it will be seen that if the arrises of the rail were required, the dotted line through the bottom corner would give the least thickness that the rail could be got out of, and the dotted lines at right angles to VT the width at the wide end. It will be seen that if the rail has to be got out of this thickness of stuff to keep the proper height, a slab will have to come off both top and bottom sides of the shank.

Fig. 2: Having cut the wreath out square to the size of face mould, allowing extra width at the wide end if required, the piece is planed true and the mould applied, and the tangent lines drawn on as shown by the dotted lines. The joints are made square to these lines. Mark the center of the piece at each end. With a bevel set to the long edge and riser side of the pitch-board, draw the line through the center square over a line on each face from this line as shown. This gives the new tangent lines and the distance the mould has to slide. The tangent lines on the face mould are held to these lines. Then mark for the stuff to be cut off, apply the mould on the other side and tack it on, working off the superfluous stuff to the lines and edge of the mould.

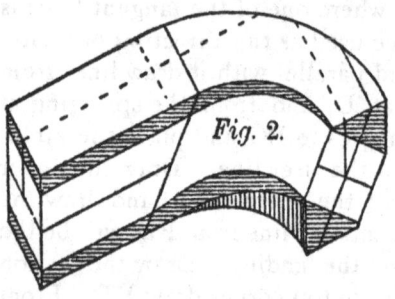

The figure represents the wreath worked into cylindrical form and ready for squaring, and the shaded portions show the slabs that have to come off. In practice this wreath is not easy to mould, owing to the rise beyond the springing on the inside.

Fig. 3 shows a better wreath. Draw the square section of the rail, and through the top corner draw the top side of rail for the lower portion, and through the bottom corner draw the under side of the top portion. Draw lines to the depth of the rail, and from the intersection of the two under sides, draw the horizontal line, and mark half the width of tread on it, measuring from the intersection. Through the point draw the perpendicular which gives the position of the risers' landing and starting; in this case they are in the

THIRD METHOD

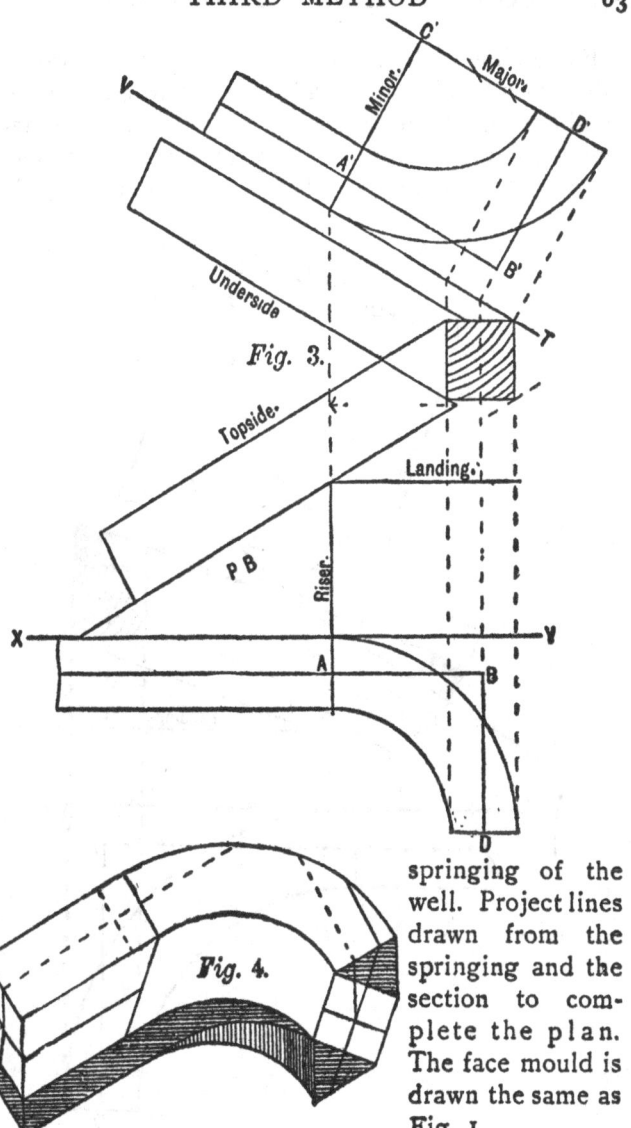

Fig. 3.

Fig. 4.

springing of the well. Project lines drawn from the springing and the section to complete the plan. The face mould is drawn the same as Fig. 1.

64   COMMON-SENSE HANDRAILING

Fig. 4 shows the wreath piece cut out and worked into cylindrical form ready for squaring, the face mould being applied the same as in Fig. 2. The face mould gives the line for squaring the top off, and a better curve is obtained. It will be seen at Fig. 3 that

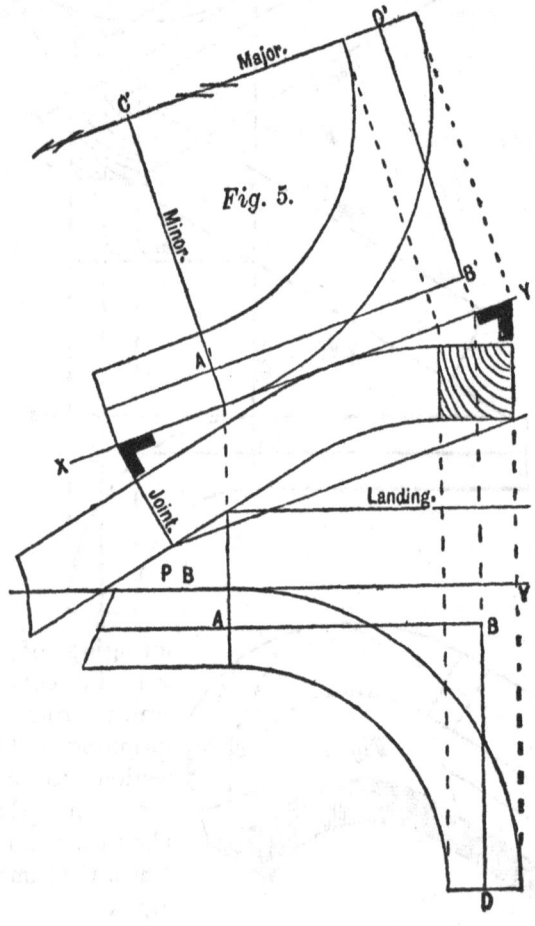

THIRD METHOD

the shank is not in the center of the stuff. The shaded portion shows the slab to come off one side.

Fig. 5 shows the face mould for a large well with the risers in the springing. Project lines from the plan and draw the section at the proper height above the landing. Draw the line from the under side of the section to meet the under side of rail coming up. Draw the line from the joint to corner of the section. This gives the inclination of the plank, and the parallel line on the top side, the thickness; the bevel at X gives the butt joint, and the bevel at Y for sliding the mould. The face mould is drawn the same as Fig. 1.

Fig. 6: The quadrant BD is the center line of rail. AB, AD are the tangent lines, the other two the springing lines, BC and CD, which are at right angles to the tangent lines, and meeting in the center C, from where the center line of rail is drawn. Then ABCD is the plan of a square prism. Two of its faces are tangent to the center line, the other two are at right angles to the tangent faces, and their intersection is the axis of the cylinder containing the center line of rail. With the side AD as a ground line draw an elevation of that face, AD, EH. Draw the line HI as the elevation of the line AD, and the pitch at which that face of the prism is cut. With A as center, turn AB into the vertical plane; draw B'I, which is the elevation of AB, and the pitch that the face is cut by. To determine the horizontal trace of the cutting plane that contains these two lines, produce HI to meet the ground line, which is one point in HT, and B being in the horizontal plane is another point. Draw HT through these points. Having the horizontal trace HT and the plan D and elevation H of a point that the plane passes through, the inclination and section

66   COMMON-SENSE HANDRAILING

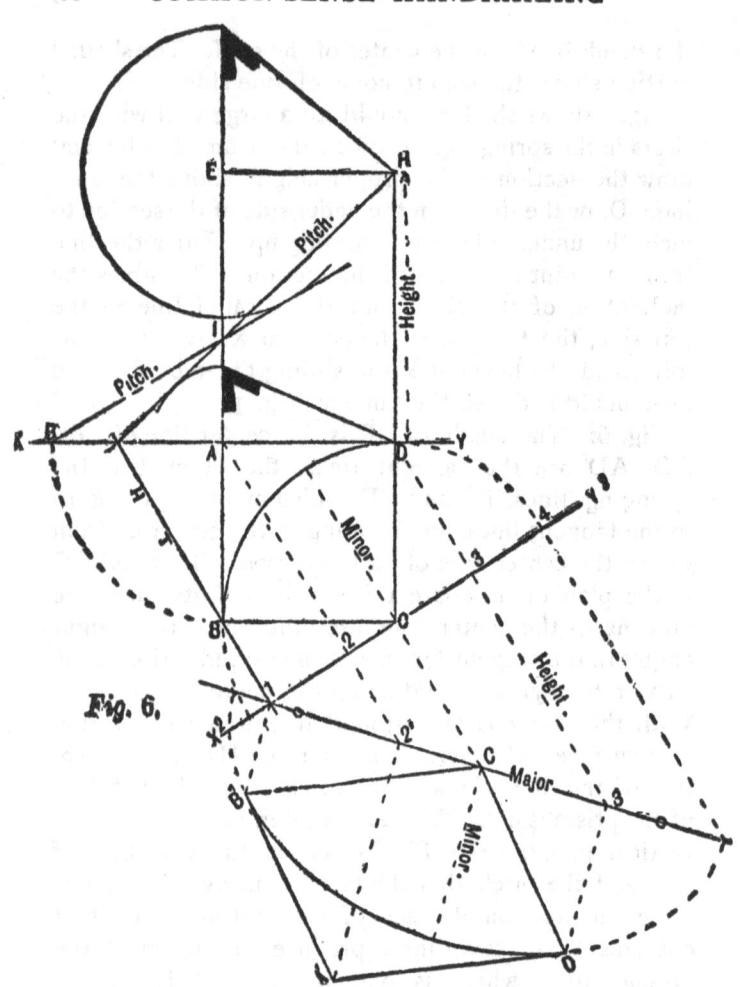

Fig. 6.

can be readily obtained. At right angles to HT draw X2, Y2 through the center C; by doing this the line that shows the true inclination will also contain the major axis of the elliptical section. Parallel to HT

THIRD METHOD

draw lines from ACD, and on the line from D set up the height 3, 3' taken from HD; then draw the line through 1, 3', which is the true inclination of the plane. From 1, 2', C', 3' draw lines at right angles to 1, 3'; then make 1B" equal to 1B, 2'A' equal to 2A, C' minor to C major, 3'D' to 3D. Join A', B", C', D'. This is the section of the prism, and to be correct A'B" must equal B'1, and A'D', HI. To draw the section of the cylinder continue the center line of rail to meet X2, Y2 at 4; draw 4, 4' parallel to 3, 3'; then from C' to 4' is the semi-major axis, the semi-minor being drawn. Draw the semi-ellipse which will pass through B" and D'.

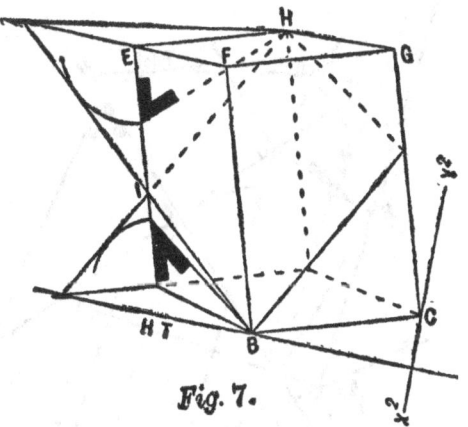

Fig. 7.

Fig. 7 is a perspective view of the prism to assist the reader to follow what is done at Fig. 1, in order to obtain the necessary bevels, that is, the angles the cutting plane makes with the vertical faces of the prism. A, B, C, D is the bottom face, and E, F, G, H the top face, and BI, HI the pitches the two faces are cut by. Produce the upper pitch to meet the ground line, AD,

68　COMMON-SENSE HANDRAILING

produced through this point and B. Draw the horizontal trace. To find the angle between the plane and this face, with A as center draw a circle tangent to the pitch-line; and to cut the perpendicular AE, draw the

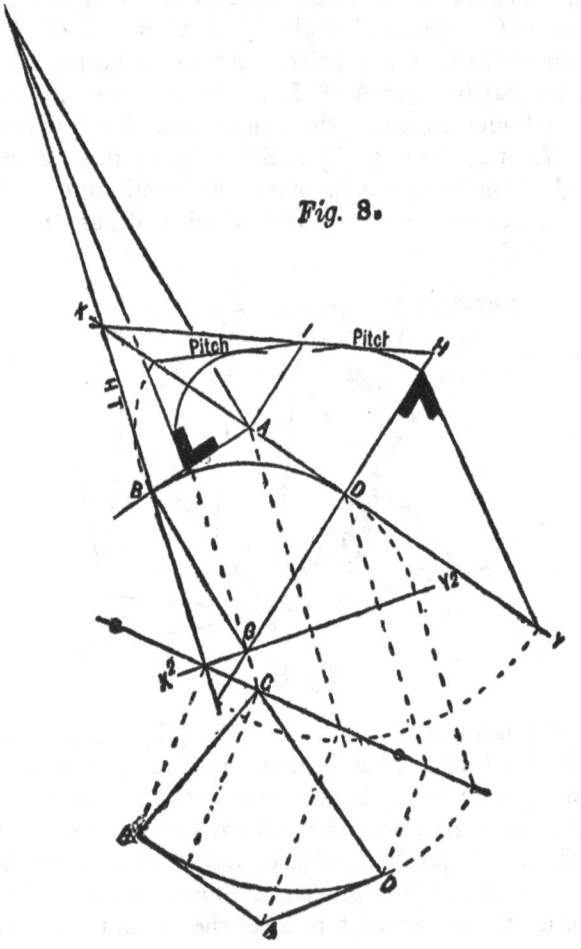

Fig. 8.

THIRD METHOD 69

line from the intersection to B, and in the angle is seen the bevel, B and D both being the same distance from A (see D, Fig. 6). Referring again to Fig. 7 it will be seen that there is a second horizontal plane containing the top face of the prism and the pitch BI produced to meet the edge. EE produced is a point in the horizontal trace in that plane, and the dotted line through H is the horizontal trace. Then with E as center draw the circle tangent BI, and cutting the perpendicular AE, draw the line from the intersection to H, and in the angle is seen the bevel between the plane and this face. It will be seen that the bevel is found the same as at Fig. 6.

Fig. 8 shows the tangent making an obtuse angle.

Fig. 9 shows the tangent making an acute angle, the construction and lettering being the same as for Fig. 6, with the exception that both bevels are found on one horizontal plane. The bevels are found on the principle of finding the inclination of an

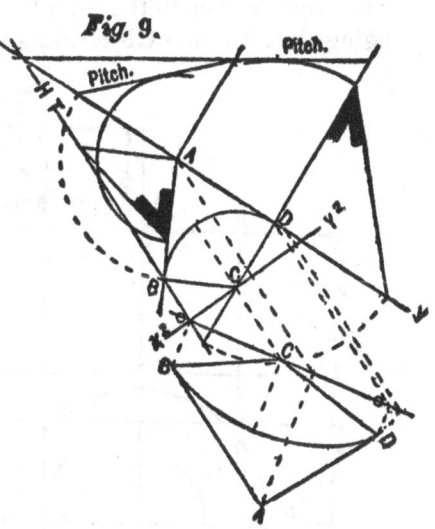

oblique plane to the vertical plane of projection, given in all books on solid geometry.

Fig. 10 is a sketch of the block given in plan at

70   COMMON-SENSE HANDRAILING

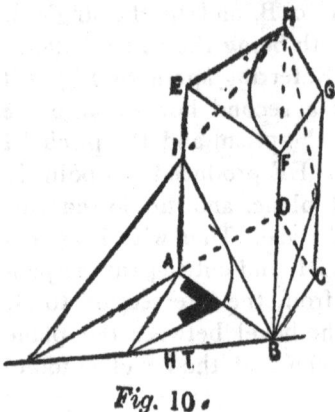

Fig. 10.

Fig. 9 to show more clearly how the bevel for the lower pitch is obtained.

Fig. 11 is the plan of a rail for a level landing, the risers landing and starting in the springing of the well. Draw the center line of rail and the joint line CD. Draw the tangent lines AB and EF, which are the center lines of the straight parts produced. The line AE is drawn at right angles to the joint line CD, and with the springing lines BC and CF forming two squares.

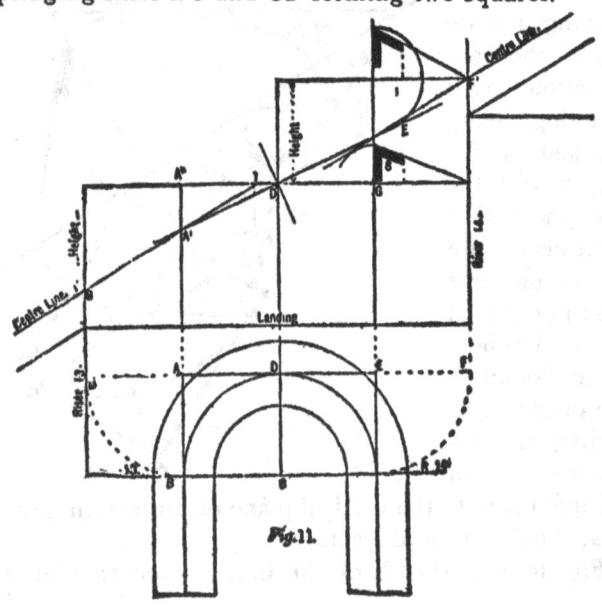

Fig. 11.

THIRD METHOD

To draw the development of these tangent lines, with A as center turn AB around, and with E as center turn EF around, erect perpendiculars from B, A, D, E, F; then place the pitch-board with the risers to the perpendicular springing line at B, and draw the under side of rail from where this cuts the perpendicular. Draw the horizontal line (marked landing) where it cuts the perpendicular. From F set up the height of a riser. Place the pitch-board, P, as shown and draw the under side of rail for the top portion. Set off half the depth of rail at both top and bottom and draw the center lines. Where they cut the perpendiculars from B and F draw the horizontal lines, and through the points where they cut the perpendiculars from A and E draw the pitch across the well. Through the point where this pitch cuts the perpendicular from D draw the horizontal line. The distance between these horizontal lines gives the height the rail rises in coming up from B to D and from D to F. In this system the heights are taken from spring-line to spring-line, the shank ends are the tangent lines produced, which may be made any length.

To draw the bevels, with G as a center on the horizontal line draw the circle tangent to the center line, and to cut the perpendicular from E and from the intersection, draw the line to H, and in the angles is seen the bevel for the shank end. Then with I as center draw the circle tangent to the pitch across the well and turn it around to cut the perpendicular above I. From the intersection draw the line to F, and in the angle is seen the bevel for the center joint. Set off half the width of rail on the horizontal lines and project them up to the bevels. Measuring along the top edge

of the bevel gives half the width of face mould at the
end the bevel is for.

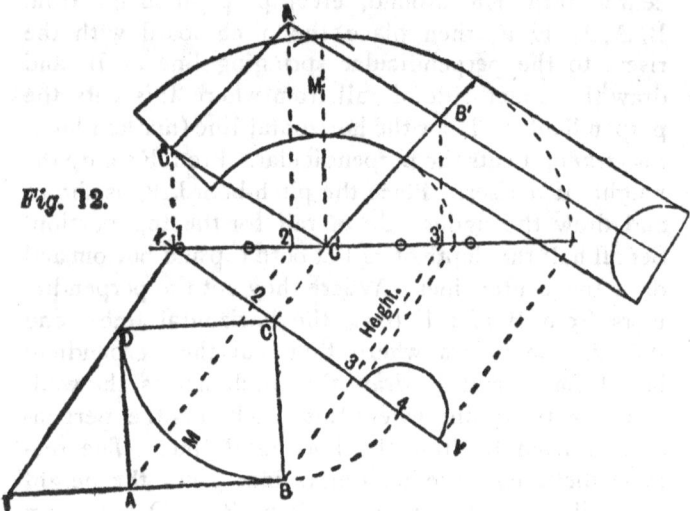

Fig. 12.

Fig. 12 is the face mould, the square, ABCD, being
drawn the same as Fig. 11. Having drawn the center
line the next step is to determine the horizontal trace
or directing ordinate, for all ordinates must be parallel
to this line.

It will be seen that the center line at Fig. 11 is
drawn through the perpendicular AA' to meet the
horizontal line at T; then A"T is the distance measured
on AB produced, at Fig. 12, that gives the point in
HT or directing ordinate, and D being in the horizontal plane is another point in the trace. Then draw
HT or ordinate, and draw XY at right angles with it
through the center C; then parallel with HT draw lines
from ABC. On the line from B set up the height 3-3'
taken from Fig. 11, draw the line through 1-3', draw

THIRD METHOD

lines from 1, 2', C', 3' at right angles to 1-3', and on these lines mark off the distances taken from the plan. Measuring from XY, 1D', 2'A'C'M'3'B', join D'A', and make the joint D' square with this line. Join A'B', and produce it to any convenient length. Make the joint square with A'B', and set off the width on each side taken from the bevel for the shank S at Fig. 1. Draw lines from D' to C' and through B'C', which is the springing line.

To draw the curve for the inside and outside of the rail, continue the center line from B to 4 in XY. Then set off half the width of the rail on each side of 4, project these to meet 1-3', as shown by the dotted lines. These two points give the semi-major axis for both curves measured from C; CM is the minor axis. Set off half the width of rail on each side of M; this gives the semi-minor axis for both curves. Draw the curves with a trammel or string and pins, which to be correct must pass through the points on the line C'B'; and the tangent lines on the face mould, B'A' and A'D', must be the same length as B'A' and A'D' in the development, Fig. 11.

In practice, as soon as the line 1-3' is drawn with all the points on it, the face mould is drawn on a thin piece of stuff with a gauge line run on at a convenient distance from the edge. This gauge line represents the line 1-3'. All the points on 1-3 are marked on the edge of the board and squared over on the face, and the distances marked from the gauge line and the face mould drawn same as Fig. 2. Then cut it out to the lines and square over on the other side the tangent, spring and minor axis lines.

Fig. 13 shows the wreath cut out square through the plank, planed true, and the mould applied, the tangent lines TT pricked off and the joints marked. The joints are made square to these lines. Square over the lines on the ends, mark the center of the stuff on these

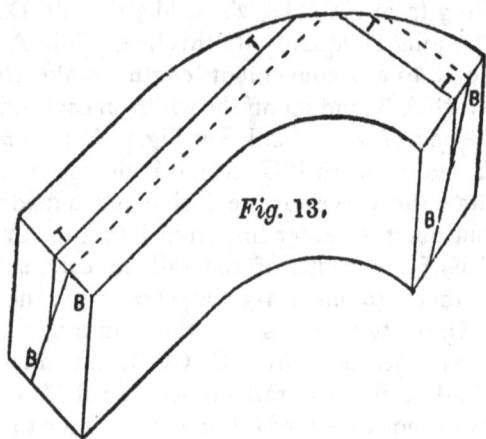

Fig. 13.

lines, then with the bevel for each end, draw the lines BB through the center of the stuff. Square over these lines on the face on both sides as seen by the dotted lines on the top side.

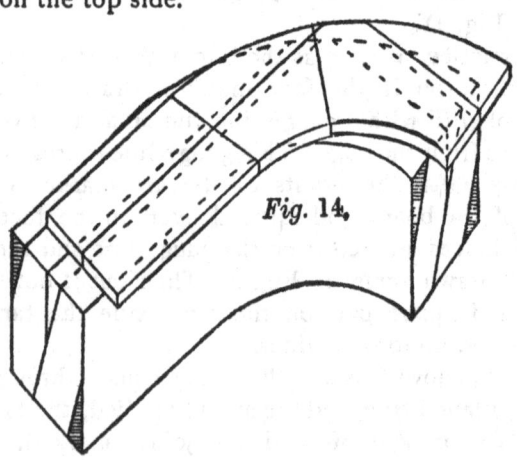

Fig. 14.

THIRD METHOD

Fig. 14 shows the face mould in position with the tangent lines held to the corresponding line on the wreath. The etched part shows the amount to be taken off.

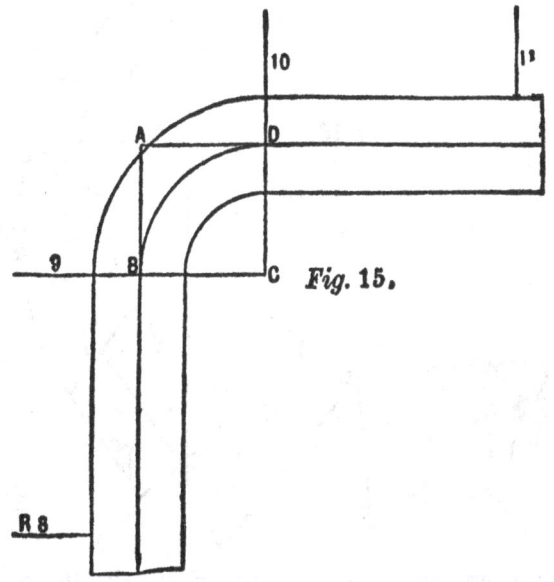

Fig. 15 is the plan of a rail for a quarter-space landing, the risers' landing and starting being placed in the springing of the well. The radius of the center line of rail being equal to half a tread, the pitch-board gives the inclination of both tangents. If a square block be cut with two of its adjacent sides to the same pitch, it will be seen that a line joining the two opposite corners is horizontal. This being known, it is unnecessary to unfold the tangents.

Fig. 16. Draw a square, ABCD, same as the square, ABCD, at Fig. 1; draw one diagonal, which is the directing ordinate; draw the line XY at right angles with it; and draw lines D and B parallel to the ordinate. Referring to Fig. 15 it will be seen that the rail in com-

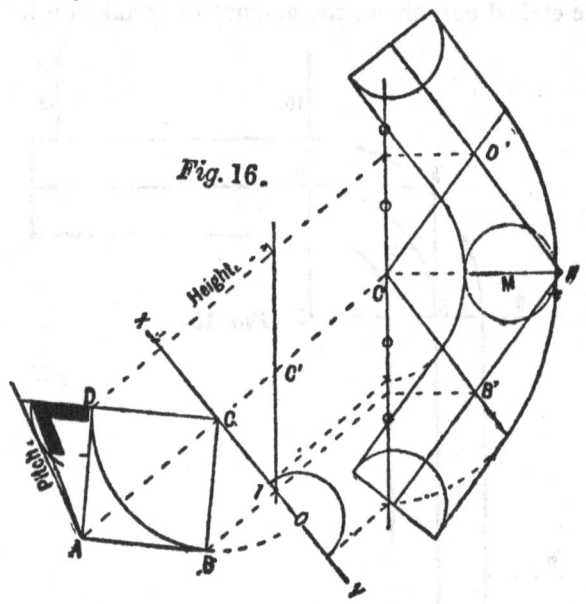

ing up from springing to springing will rise the height of one riser. Again at Fig. 16, on the line from D, set up the height of one riser. Draw the line from 1 through C; a second line is drawn parallel with the first. This line would represent the gauge line on the stuff for the face mould. Square over the lines on the stuff; from 1, C,'3', on these lines, mark off the distances taken from the plan 1B', CMA, and 3D, and through the points draw the tangent lines A'B', A'D', and B'C', C'D' the springing lines. Produce the tangent lines any convenient length and make the joint lines square with them.

To draw the bevel, place the pitch-board on the side

THIRD METHOD

AD of the square with its tread side to the line, and draw the pitch through A, which is the inclination of the tangents. Then with D as center draw the circle tangent to the pitch, and cutting the side CD produced, draw the line from the intersection to A, and in the angles is seen the bevel for both ends. Set off half the width of rail on DA, project it up to the bevel. This gives on the top edge half the width of face mould at each end. Set off this width on each side of the tangent lines A'B', A'D', and draw parallel lines through the points to the springing lines C'B', C'D'. To draw the curves turn the center lines around to meet XY at O; on each side of O set off half the width of rail; project these on to the line 1-3', which gives the semi-major axis for each curve. On each side of

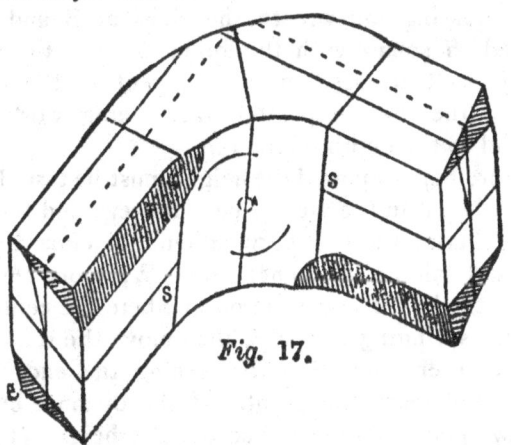

Fig. 17.

M on the minor axis line, set off half the width of rail; these give the semi-minor axis for each curve. Draw the curves which must pass through the points on the springing lines on each side of B' and D'.

Fig. 17 is the wreath worked into cylindrical form,

and the lines shown ready for squaring. To draw the lines on the stuff, a thin piece is made to the size of the rail before being moulded, and a gauge line run through the center both ways. The one through the depth is held to the line drawn across the end or joint with the bevel, and the other kept to the center of the stuff. Mark the top and bottom, which gives the shaded portions to come off. The line through the center is squared over on each side as far as the springing lines at SS, the minor axis line is drawn across the stuff parallel with the springing lines. Mark the center of the stuff on this line, and with O as a center, and the compasses set to half the depth of the rail, draw the two arcs as shown. The outside of the wreath is marked in the same manner.

In squaring the wreath the slabs at B and T are worked off square with the joints as far as the springing lines, then eased around tangent to the arcs of circles, the two opposite slabs being worked off parallel to the depth of the rail.

Before being squared the height must be tested. The wreath is put in the vice or bench screw, and the sides and springing lines set perpendicular to a board planed true and laid on the bench top. With one end of a rod on the board, mark the point where the center line cuts the springing line at S, then move the rod around to the other point S, still keeping the end on the board, and mark the point. If the distance between the two points on the rod equals the height at Fig. 2 the wreath can be squared, if not, the center lines must be raised or lowered to suit.

Fig. 18 is the plan and stretchout of the tangents for another quarter-space landing, the center line being struck with a smaller radius than Fig. 1. The tan-

THIRD METHOD

gents are unfolded and the center line drawn at the top and bottom portions as shown. The center line

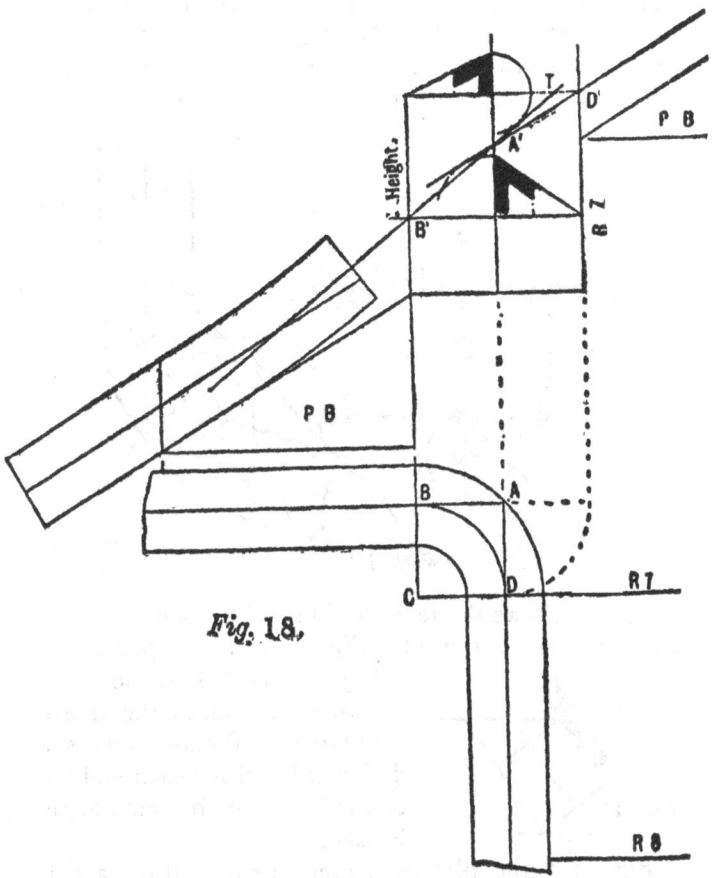

Fig. 18.

A'D' is produced past A', which is the point where the center line of the lower portions must meet it. It will be seen that if the center line of the lower portion had been produced, it would cut the perpendicular below

A'. Then the pitch must be altered to lengthen as few balusters as possible. This must be done from about the center of the last step, as shown by the pitch-board. Draw the line B'A' as shown. Bisect the angle; draw the joint-line to meet the bisecting line, which is the center the ramp is struck from.

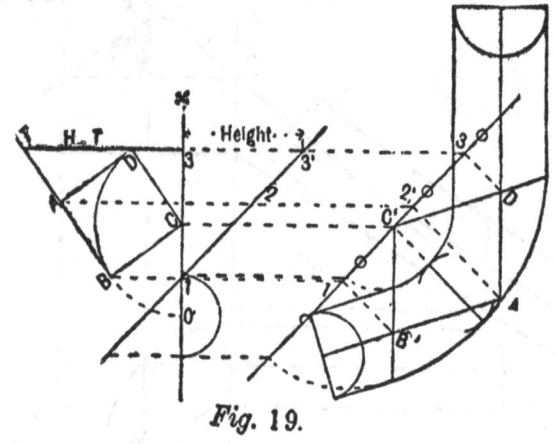

Fig. 19.

Fig. 19 shows the face mould for Fig. 18, obtained in the same way as those for Fig. 1 on the last page.

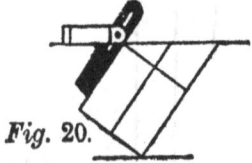

Fig. 20.

Fig. 20 shows how the usual thickness allowed for the wreath is obtained. Where there is a different bevel for each end the one with the more acute angle is used.

Fig. 21 is the plan of center line of rail for a well with six winders, as shown, two of them being in the springing at each side. Draw the tangent lines, the joint line CD, and the springing lines BC, CF forming two squares as shown.

THIRD METHOD

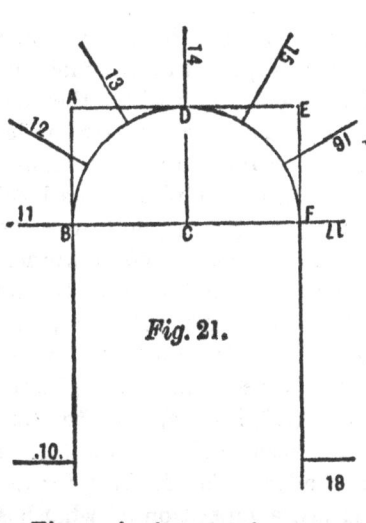

Fig. 21.

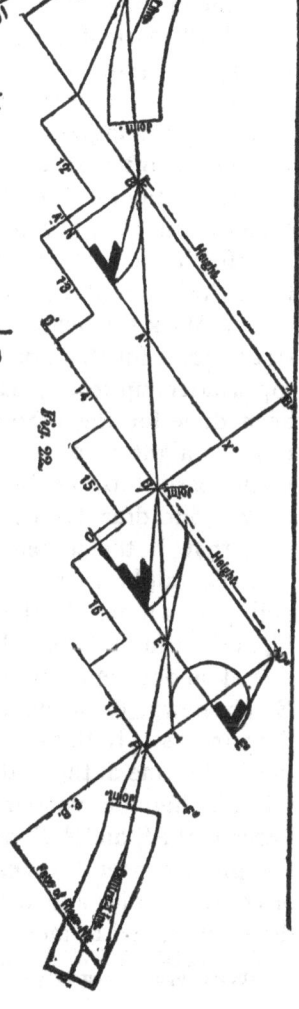

Fig. 22 is the stretchout or development of the tangent lines. To draw this let the line LM be the edge of the drawing board; with a bevel set to a convenient angle the end of the winder and riser would give the pitch. Draw the line B'B"; then parallel with B'B" draw A'A" at a distance equal to the side BA of the square at Fig. 21, and the same with D'D", E'E" and F'F". Place the pitchboard with its riser side to the line B'B", as shown, and draw the last straight step. From the top of the riser draw the line at right angles to B'B" for the first winder. This being the development of the tangents the width of each winder is taken from where they cut the tangent lines at Fig. 21. Then on the first winder from B'B" mark off the distance B to R12*, Fig. 21, and through the point R12', with the pitch-board mark off the height of a riser; and through the point draw the line at right angles to A'X" for the second winder. From A'X" mark off the distance A to R13, Fig. 21, and through the point draw R13' the height of a riser, and draw the line at right angles to D'D" for the third winder.

Referring to Fig. 21 it will be seen that riser 14 passes through the point D, then R14', Fig. 22, is on the line D'D". Mark the height of a riser, and draw the line at right angles for the next winder. The drawing for the top winders up to R17' is just a repetition of what has been done for the others. Then from R17' draw the next straight step, and from the top edge of the pitch-board, top and bottom, set off half the depth of the rail and draw the center line as shown. It will be seen that, if the center line at the top had been continued straight to meet E'E", the rail would be too high over winder 16; then the rail must be lowered at F', but not more than half its depth. From the center line at about the center of the step, draw the pitch to E', as shown; draw the pitch A'E' nearly parallel with the winders. In this case it is continued down to meet the center line at the bottom. If this should make the rail too high at winder 13, then the pitch must be lowered at A', and A'B' drawn to another pitch. Where the pitches meet the center lines at top and bottom, bisect the angles and draw the joint lines square with the pitches, and to meet the bisecting lines which give

---
*R. 12 means riser 12 on Fig. 21, and R. 12', riser 12' in Fig. 22; and so on with these compound references.—*Ed.*

THIRD METHOD

the centers the ramps are struck from. The joints should be kept clear off the springing lines B'B" and F'F". The ramps may be made any length. The center lines must be marked on the templets and the face of the risers 10 and 12, as shown, and transferred to the ramps. The lengths of the straight rail can be got from these riser lines, and all can be jointed with accuracy.

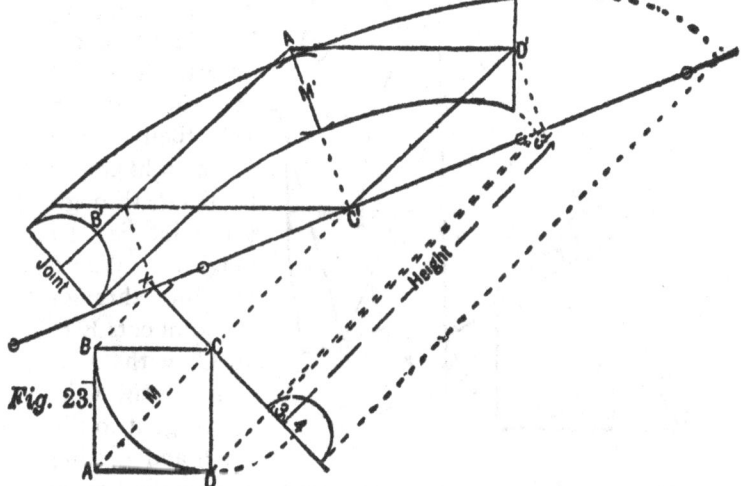

Fig. 23.

Fig. 23 is the face mould for the lower portion, the drawing being similar in every respect to what has already been described. The height is taken from where the pitch or raking tangent cuts the springing line B'B", Fig. 22, to the line drawn at right angles to D'D" through the joint. The pitches being the same, the bevel answers for both ends. Where the line from B' cuts A'A" at N, with N as center, draw the circle tangent to the pitch and cutting A'A". From the intersection draw the line to B', and in the angle is seen the bevel. The straight portion of the face mould from B′ to joint at Fig. 23 is made the same as B′ to joint at Fig. 22.

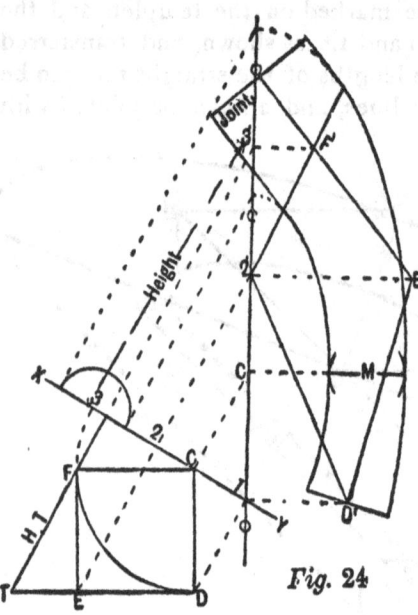

Fig. 24

Fig. 24 is the face mould for the top portion. To find the height where the pitch cuts the springing line F′F″, Fig. 22, draw the line F′F″ at right angles to D″; then to the line at right angles to D′D″, through the joint is the height. Where the line through the joint cuts E′F′ at O, with O as center, draw a circle tangent to the pitch and cutting the line E′E″. From the intersection draw the line to D′, and in the angle is seen the bevel for the shank end; where the line F′D″ cuts E′E″ as center draw a circle tangent to the lower pitch, and cutting the line E′E″. From the intersection draw the line to D″, and in the angle is seen the bevel for the center joint. Make F joint, Fig. 24, same as F joint, Fig. 22.

Fig. 25 shows a simple method of finding the bevels, and one which answers in every case. Let LM be the edge of the drawing board. Square over the line CM, and make the line CM equal to the radius of

THIRD METHOD

the center line of rail. Then at Fig. 24, with C as a center, draw an arc tangent to the line E'D'; then with C, Fig. 25, as a center and the same radius, draw the arc cutting the edge of the board, as shown; draw from the intersection to C, and in the angle is the bevel for that end of the wreath. Again at Fig. 24, with C' as a center, draw an arc tangent to the line E'F', and repeat at Fig. 25 for the bevel for that end of the wreath. These bevels can be tested with those at Fig. 22, and they will be found to be exactly the same.

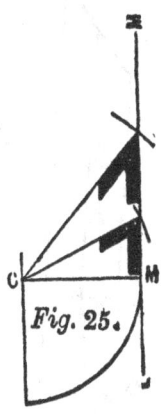

Fig. 25.

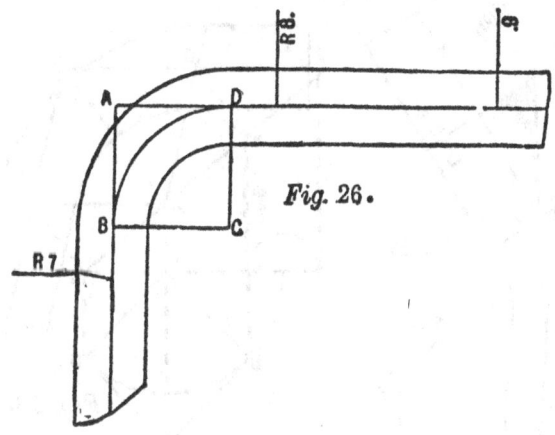

Fig. 26.

Fig. 26 is the plan of the wreath, showing the risers and tangents.

Fig. 27 is the development, being a repetition of what has been done in previous chapters. It will be seen that the center lines of the straight rails (if pro-

86 COMMON-SENSE HANDRAILING

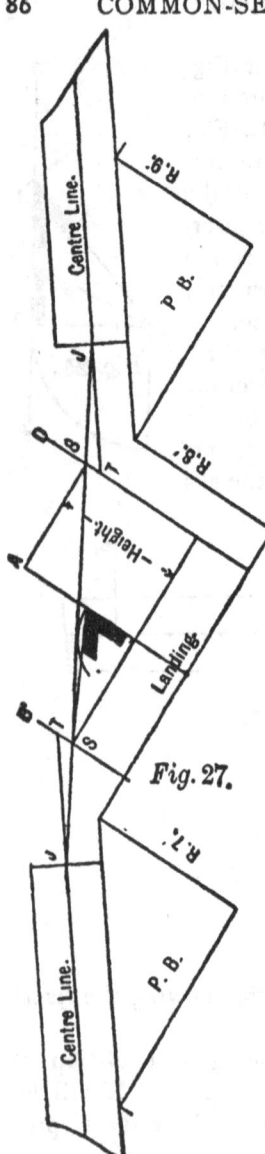

Fig. 27.

duced) would not meet on the perpendicular A'. Then there are three ways this wreath may be worked out by the tangent system: first, with the wreath in one piece, and to form its own easings. To do this settle on the position of the joints JJ (the shorter the shank ends the less thickness will be required for the wreath); from JJ draw the pitch across the well. The point where this line cuts the perpendiculars B' and D', gives the height.

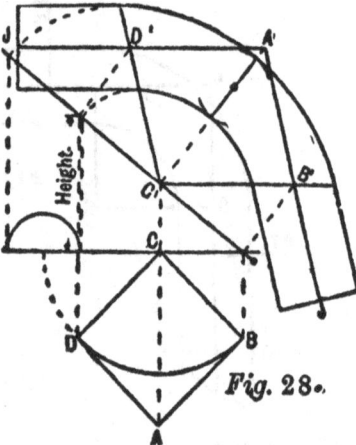

Fig. 28.

Fig. 28: The pitches of the wreath being equal, the face mould is drawn same as those in previous plates.

THIRD METHOD

Fig. 29 is an isometrical sketch of the wreath worked into cylindrical form. Cut out the wreath a little wider than the face mould; square through the plank; plane one face true and apply the face mould, marking the joints and transferring the tangent lines to the stuff. Those are represented by the dotted lines. At first the joints are made square to those lines and to the face of the wreath. Square the lines over the ends as far as the center C; then with the bevel, shown at Fig. 27, draw lines across the ends through C. From those lines, draw lines square from the end on each face of the wreath. The tangent lines on the face mould are held to these when marking and working the wreath into cylindrical form. When the face mould is in its position on the wreath (when worked), draw lines across each side of the wreath from the springing and minor axis lines as shown. Draw the line through C square with the line drawn with the bevel. From this line draw square from the end the center line JS on the side of the wreath. Referring again to Fig. 27 it will be seen the center line of the straight rail meets the perpendicular springing line D' at T. Then make a thin piece of stuff to the exact shape of the triangle JST and apply this to the wreath at Fig. 29. With the side JS to JS, and ST to the springing line draw

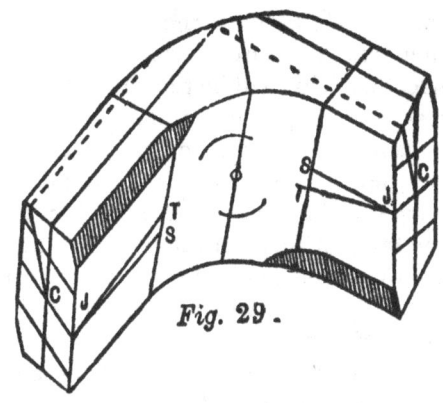

Fig. 29.

the line JT, which is the new center line. The joint is then made square to this line and the side of the wreath. This will be repeated at the other end. Then test the height and square the wreath. The shank ends must be worked off as far as necessary parallel to the new center line JT. Should it be necessary to have the shank ends the exact length shown at Fig. 27 the tangent lines on the face mould, B'J and D'J, Fig. 28, would be made to TJ instead of SJ, Fig. 27, to allow TJ on the wreath, Fig. 29, to be the same length as TJ, Fig. 27. This would be immaterial in joining up to straight rails, as the springing lines are drawn across the under side of the wreath and the lengths

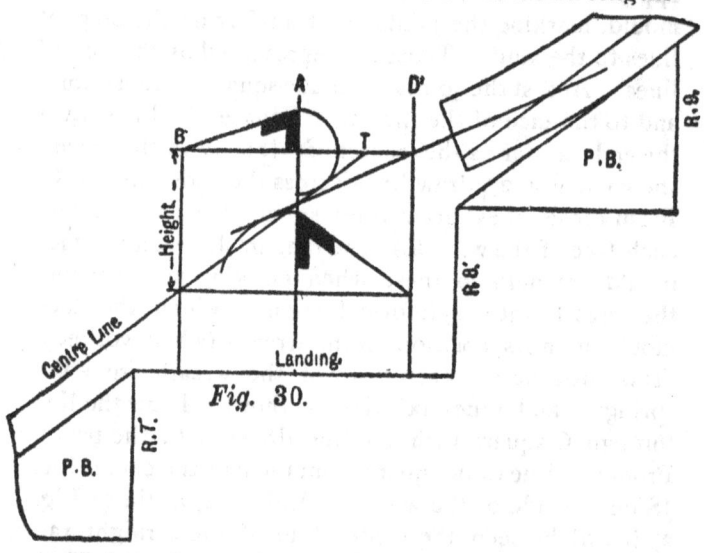

Fig. 30.

are taken from them. For the extra thickness required for this wreath the distance ST, Fig. 27, would be quite sufficient over what is usually allowed.

Fig. 30 is the development for the wreath to be in

THIRD METHOD

one piece with the easing on the straight rail at the top. The face mould is not drawn, being similar to those already explained.

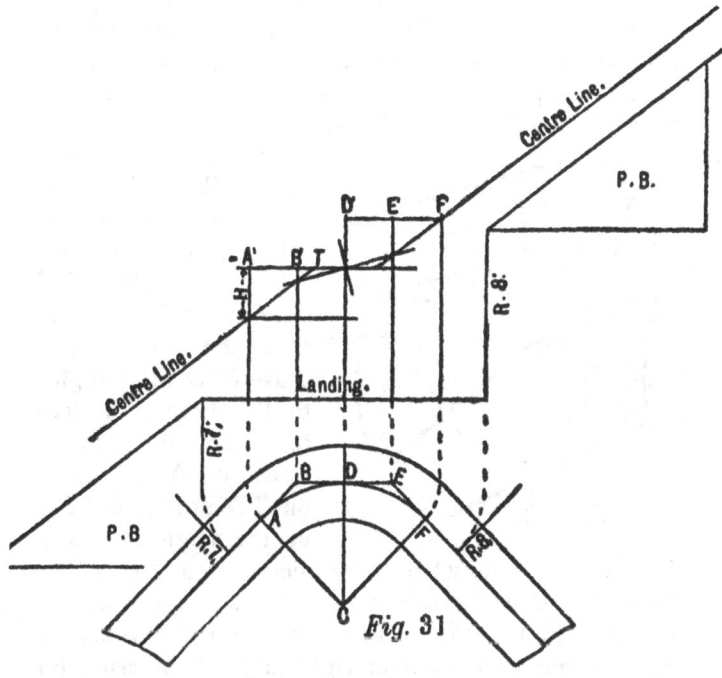

Fig. 31

Fig. 31 is the development for the wreath to be in two pieces. Redraw the plan same as Fig. 1; draw the joint line CD, and draw BE tangent to the center line and at right angles to CD. Produce the center lines past A and F to meet B and E. With B and E as centers turn the springing lines around, and project them up parallel with CD. Draw the landing line at right angles to CD; place the pitch-board at R7 and draw the pitch and center line. At R8 set up one riser above the landing; then draw the pitch and center lines. From where the center lines cut the perpendiculars B' and E', draw the pitch in the center of the wreath, and where this cuts the perpendicular D' gives one point in the height, and the joint is drawn through the intersection, and where the center lines cut the perpendiculars, A' and F', gives the other points in the heights, as shown by the horizontal lines.

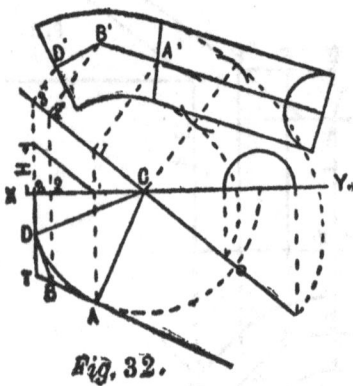

Fig. 32.

Fig. 32 is the face mould. Redraw the center line with its tangents AB and BD; produce AB to T, making BT same as B'T, Fig. 6. Draw the ordinate TD; draw XY at right angles to TD through the center C and draw projections from A and B, and on TD produced. Set up the height H, taken from Fig. 6, from where the projector from A cuts XY. Draw the pitch of plank and parallel with this draw the line through C. Where the lines from TD, B and A cut this line draw lines at right angles. Measure on these the distances 1A, 2B and 3D, join A'B'D' and make the joint square with B'D'. Draw the springing line through A' to C and on the line from C set off the radii of the inside and outside of the rail for the minor axis. Turn the center line around to meet XY at O, set off half the width of rail on each side, project these down to meet the pitch-line for the major axis, and draw the curves.

Fig. 33 shows the bevels. Let LM be the edge of a

THIRD METHOD

board. Square over a line and measure off on this the radius of the center line of rail CA. Then with C, Fig. 32, as center, draw an arc tangent to A'B'. Then from C draw the arc, cutting LM; draw from the intersection to C, and in the angle is the bevel for the shank end. Set off half the width of radii on AC; project this up to the top edge of the bevel; take the distance along the top edge of the bevel and set it off on each side of the tangent A'B', Fig. 32. Draw the lines parallel with A'B to meet the springing line. The elliptic curves must pass through these points. The shank end can be made any length. For the bevel at the other end the arc must be drawn tangent to B'D', Fig. 32, and repeated as shown.

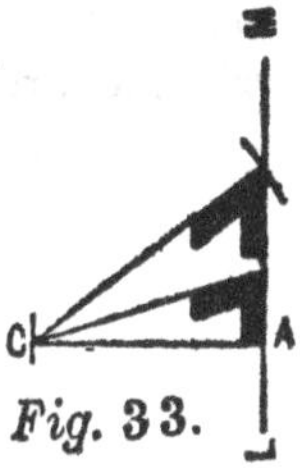

Fig. 33.

In the system of handrailing, known as the section of a cylinder, through three given points, or the face mould plane through three points, the section is determined through an imaginary solid containing on its surface the center line of rail, and its base being defined by the plan of the center line. The following examples illustrate this method:

Fig. 34 is the plan and development of the center line of a rail for a well with two quarter-space landings, the risers being placed in the springing at each side of the well. To draw the development of the center line, draw the equilateral triangle on the diameter of the center line. Produce the two sides to cut the line drawn tangent to the center line and parallel with the diameter. Between the intersections is the stretchout or development, and the perpendicular lines S, S are springing lines. Place the pitch-board with its riser side to the line S at R7; draw the line along the top edge where it cuts the springing line; draw the first landing at right angles to the springing

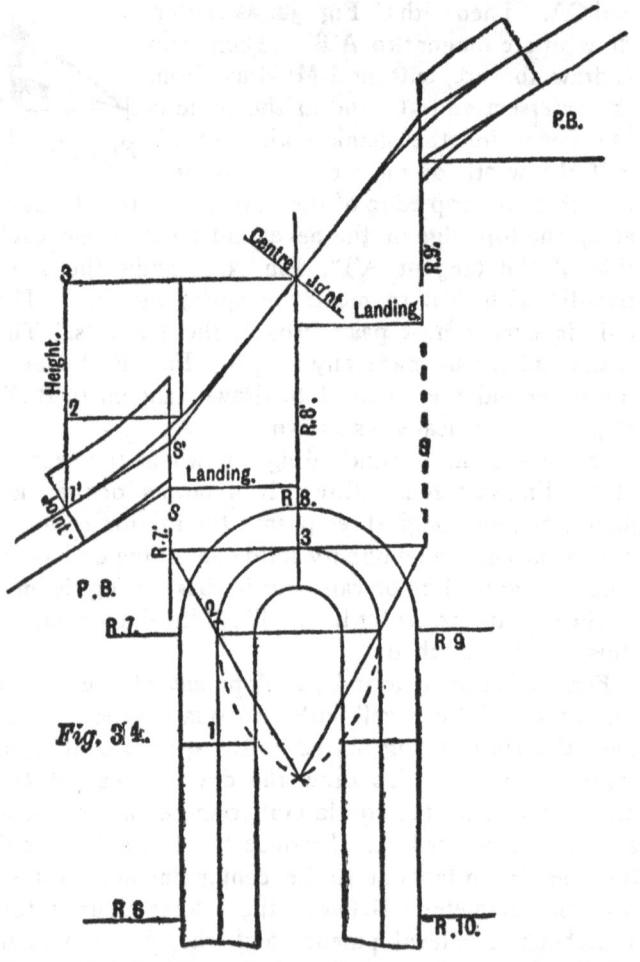

Fig. 34.

line. On the line projected from the center of the well, mark off the height of a riser and draw the second

THIRD METHOD

landing; on the springing line mark off the height of a riser and draw the horizontal line. Place the pitchboard with its tread side to this line, and the point to the springing line. Draw the line along the top edge for the under side of rail and from the top edge of the pitch-board at top and bottom portions, set off half the depth of rail, and draw the center line to meet the springing lines at S'S'. Join S'S' and draw the easings between the two lines at top and bottom, and where the line S'S' cuts the perpendicular from the center of the well, draw the joint line through the intersection, also the horizontal line to the left. Draw the joint line at the shank of the lower portion at right angles to the center line. From the center of the rail erect the perpendicular 1'3' to meet the horizontal line through the center joint. 1'3' is the height the bottom portion of the wreath risers, and also the height of two of the points the section plane must pass through. The middle resting point may be taken in the center of the development for that portion if the shank is short, as in this case, but when the shank is long the middle resting point must be taken in the curve, about one-third the distance between the springing line and the center joint. In this case the horizontal line through the center joint and 3' is divided in two and the perpendicular dropped to meet the easing line from the intersection. Draw the horizontal line to meet the perpendicular at 2, which gives the height of the middle resting point the plane must pass through.

To draw the section these points must be determined in plan. From 1', in the center of the shank joint, draw the horizontal line to the springing line. Take this distance in the compasses and mark the center line in the plan from springing to the joint at 1. Take the distance from the springing line to the middle point at the easing and mark the distance from the springing to 2 in the plan; then 3 at the center joint in plan gives the third point.

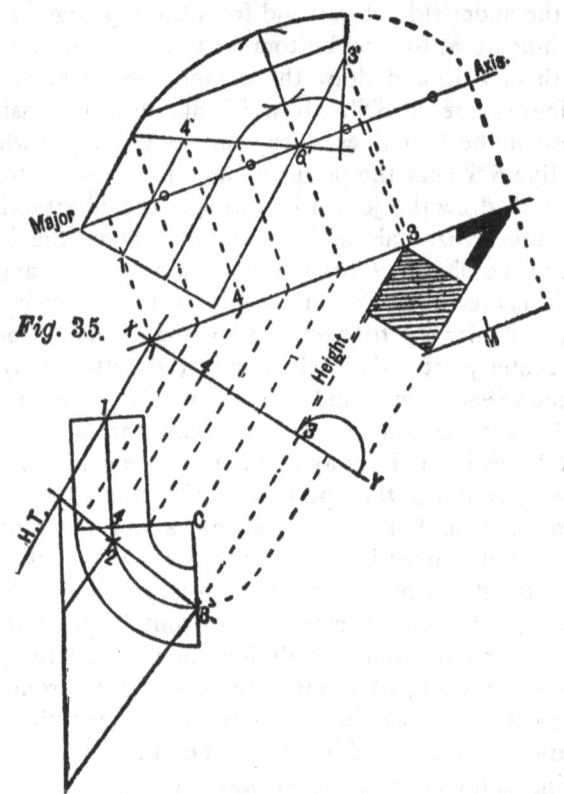

Fig. 35: Redraw the plan with the points as shown. Join 3 and 2, and produce the line to the left and on this line erect perpendiculars from 3 and 2. Make the one from 3 the height of 1'-3' at Fig. 34, and the one from 2 the height of 1'-2', Fig. 34. Draw

THIRD METHOD

a line through the points meeting the line joining 2 and 3, which is one point in the horizontal trace. The point 1 at the center of the shank joint is in the horizontal plane. Draw HT through these points; draw XY at right angles to HT; draw ordinates from the center and each side of the rail at the springing line. Across the rail through C and from 3, at the center joint on the line from 3, set up the height, 1'-3', Fig. 34; above XY draw the lines 1, 3', which is the pitch-line. From where the ordinates cut this line, draw lines at right angles and measure off on these lines from the pitch-line, the corresponding points in plan, measured from XY. The curves may be drawn with the trammel. Turn the center line around and draw the line tangent to it parallel with the ordinates. Set off half the width of rail on each side of this line and project these up to 1-3' and to the major axis line, which gives the semi-major axis for both curves, the semi-minor being on the line. Through C, draw the springing line 4'C', and draw the line through 3'C'. The joints are made at right angles to these produced at 1' and 3', as shown. Draw the shaded section, and the line through the bottom corner shows the thickness for the square rail. In the angle is seen the bevel to slide the mould by. The dotted line from the top point of the bevel cutting the line through the bottom corner of the section gives the whole distance the mould has to slide. Bisect this line at M and draw the line on the face mould through the center of the rail on the minor axis line parallel to the major axis and pitch-line. Apply the face mould to the plank, and cut the piece square through, a little wider than the face mould. Plane one face true and transfer the minor axis line and the line parallel with

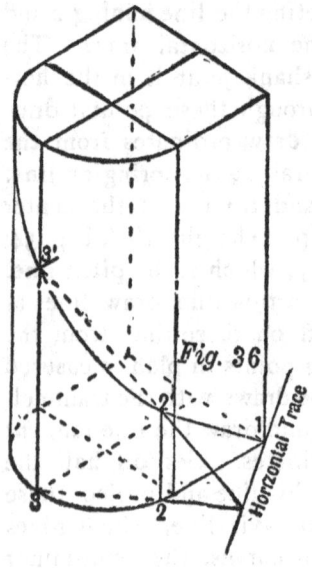

the major axis from the mould to the stuff. Square over the minor axis line on both edges and mark the distance from M to the dotted line at Fig. 35, on one side of the minor axis line at the top and on the other at the bottom.

Fig. 36 shows an isometric projection of the cylinder with the lines on the surface.

Fig. 37 shows the face mould in its position when held to the tangent lines. This is known as the square cut.

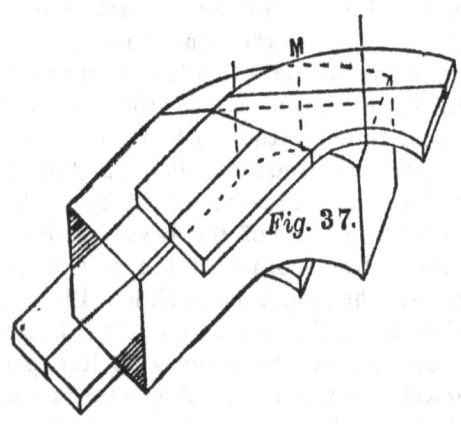

Fig. 38 shows the mould marked on the top and bottom faces of the plank, the mould being moved along

THIRD METHOD 97

the line on the under side to the distance given at Fig. 35. The wreath is cut out to those lines. This is

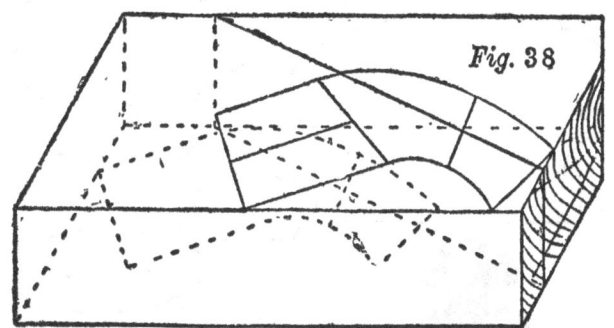

Fig. 38

known as the bevel cut. The top portion of the wreath is the same as the bottom. To make the joint at the shank end, it may be necessary to use the short piece of falling mould as at Fig. 34.

This last example is after the system invented by Peter Nicholson, and is, in fact, the foundation of all scientific methods of handrailing, though the system has been very much improved by modern handrailers.

As this little book is intended only for instruction in handrailing, very little has been said regarding the construction of the carcasses of stairs themselves, that subject being left for future consideration, as it was not considered wise to overload this volume with matter not pertinent to the subject in hand, and thus increase its selling price beyond what the workman would care to pay.

It is hoped that out of the numerous examples given, the searcher for instruction in handrailing will find more than enough to compensate him for the outlay of money and time he will expend on this little volume.

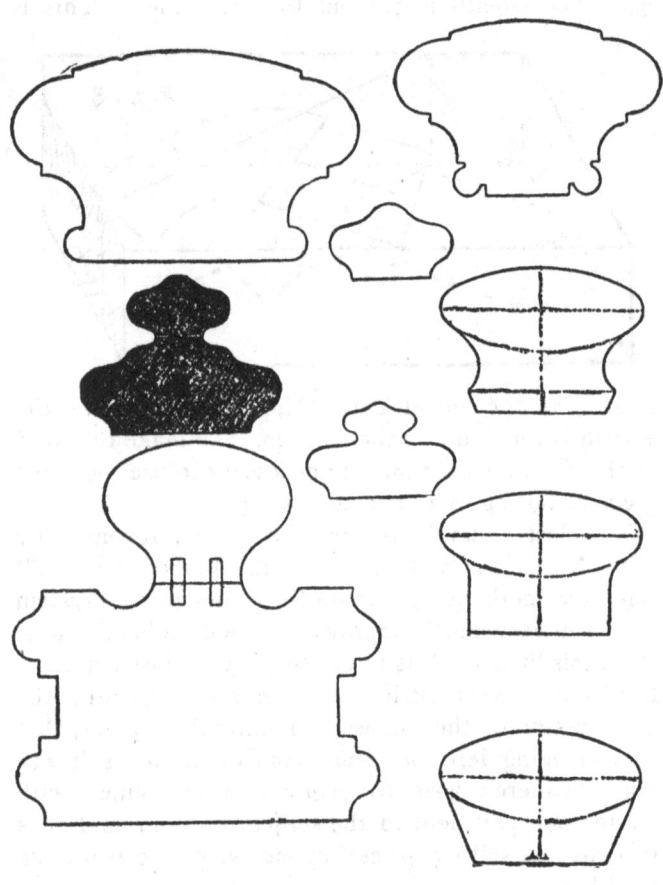

SECTIONS OF HANDRAILS

INTRODUCTION TO METHOD IV

NEWELLED OR PLATFORM STAIRS

With the introduction of the so-called Queen-Anne and Eastlike styles of building some thirty years ago, the newelled or platform stairs came more and more in vogue, and at the present time more than half the stairs that are erected are of this kind; and this fact has, in a great measure, done away with the necessity of a study of the science of handrailing by every workman who aspires to be a stairbuilder and handrailer. But while that necessity is removed to a large extent, the ambitious young workman should make a successful attempt to master the art of circular handrailing, as it will open up beauties to his mind he never could have appreciated otherwise, and will broaden his knowledge, and enable him to deal with knotty questions of joinery with skill and speed. Platform stairs are easy to construct when once the plan is determined, as newels are placed at the angles, thus doing away with sweeps and curves in the rail, or bending of the strings. They are cheaper than stairs having circular strings, and may be made to have a handsome and impressive appearance. The newels and balusters can assume almost any size and style. The stairs may have open strings, or closed ones to suit the style of architecture. Newels may be massive or slight, "built-up" or made of one solid piece, as may be desired; but where the newels are large, I would advise they be "built up," as a solid newel is likely to check and split and get out of shape.

Stairs of the kind under discussion can be made an attractive feature in a house. Every architect knows this; but no man can build a flight that will be comfortable, or even safe, in a cramped or narrow hall. Stairs are exacting in their demands, and if these demands are not complied with we shall be reminded of the neglect every time we use them. We may resort to make-shifts (if inclined to do so) in other parts of the house, but we cannot put off the stairs with anything and say "it will do," and no coaxing will bring an ill-contrived or badly-arranged flight of stairs into use on any possible terms. A good run is what every flight of stairs requires. If the run is not long enough, then we must increase the height of the risers; and the rise, after it has reached a certain point, becomes trying, then difficult, and at last dangerous. In many houses, in almost all cheap houses, the rise is eight inches. Even the back stairs should not have a more rapid rise, and for the principal stairs this is wholly inadmissable. The other extreme, a fault not often committed, is to have the rise too low.

There are great varieties of rise given to stairs for various purposes, and rules have been laid down for calculating the proportion of tread to riser. A modern writer has given seven different proportions adapted for buildings of different classes. His most ample tread is 12 in. with a 5½-in. riser; his next, 11½ in. and 5¾ in.; then follow 11 in. and 6 in., 10½ in. and 6¼ in., 10 in. and 6½ in., 9½ in. and 6¾ in., concluding with 9 in. and 7 in. We may say that a 9-in. tread is about the least that is usually allowed in practice when there is any attempt made to study ordinary comfort, although we have met with 8-in. risers and 8-in. treads in suburban villas, which, of course, gives

FOURTH METHOD

an angle of ascent of 45 deg.; while in the seven foregoing proportions this angle varies between 24 deg. and 37 deg. It is often expedient, however, to make it lower than 24 deg. With regard to rules for calculating the proportions of steps, some persons maintain that the tread and riser added together should equal 18 in. This would give 13 in. and 5 in., 12 in. and 6 in., 10 in. and 8 in., and 9 in. and 9 in., and in the two latter proportions the rise is too great. Others say that the tread and riser multiplied together should equal $17\frac{1}{2}$ in., which will give 13 in. and about 5 in., 12 in. and $5\frac{1}{2}$ in., 10 in. and $6\frac{3}{8}$ in., 9 in. and $7\frac{1}{3}$ in., and 8 in. and $8\frac{1}{4}$ in. This rule gives better results than the former. Whether the risers are high or low, they must all be of a uniform height. Any departure from this rule is always attended with mischievous results. If all the risers in a flight are seven inches, with one exception, and that one is either six or eight inches high, every person who passes up or down will trip at that step. No matter how often he goes up or down, he will always trip at that point.

The practical difficulties in arranging stairs to rise from one level to another with a sufficient tread and a commodious headway are often great, while in construction awkward problems are frequently suggested from the necessity of carrying flights of stairs over spaces where they can neither be well fastened into the side walls nor supported from below. Not only do these practical difficulties have to be considered in every class of staircase, from that of a cottage to that of a palace, but in all situations where the stairs form a conspicuous feature and where there is any pretense at ornamental building, its artistic treatment affords ample scope for the skill of the architect or the workman. Stairs of this kind, to be effective, should be wide between the wall and rail, with one or two flats or landings. The rail must be heavy, the balusters something more than "broom handles," and at the foot let there be a newel, on which the architect may display his taste and skill. It need not be elaborate, but it is a conspicuous object, and it should have something more to recommend it to our notice than the cheap and stereotyped forms, which may be bought at the turner's by the hundred. As a first and most essential principle, a staircase should present an inviting aspect, suggestive of an easy ascent, not of a painful and laborious effort at climbing. Therefore, even if it were, as a rule, possible, which it rarely is, to arrange several flights in a direct line, it would be undesirable to do so; for, however imposing the effect of such an arrangement, it could not but oppress those about to ascend it with an uncomfortable sense of coming fatigue, suggested by the prospect of one long ascent, broken only by landings which would be lost to view from the bottom.

It is pleasant to mount up stairs properly planned, especially if they are well lighted and ventilated. And if on the first landing the architect can contrive a bay, deeply recessed and provided with seats beneath the wide windows, he will, by so doing, add another charm to the house. Here, those who are advanced in years, and who find it difficult to climb one flight at a time, may rest awhile, or sit and chat. Here the little ones love to pause in their passage up and down, and here flowers growing in a *jardiniere* in front of the window, may send their fragrance through the house.

Stairs may be of wood, stone, marble, brick, terra cotta, iron, or iron and concrete. The arrangement

FOURTH METHOD

and construction of staircases forms one of the most important, and often most difficult branches of architecture and building.

Modern stone steps are either solid or formed with treads and risers. It was the latter mode of construction that probably first suggested the nosing which is found in the buildings of the Italian renaissance, erected during the sixteenth century. It is quite clear that stone stairs of the tread-and-riser construction require firm support at each end, and it is for this reason that they are seldom used except in basements. Most stairs, whether in stone or marble, are usually solid, and depend for support upon being tailed into the wall at one end, and being connected together with bird's-mouth joints, by which means each step is sustained in position by the one immediately below it, so that the thrust of an entire flight is transferred from top to bottom. In wide flights—those exceeding 4 ft. in width—it is often expedient to strengthen the connection of the stairs by means of a flat bar of rolled iron fixed to their ends with small bolts let into the stone and run with lead. Sometimes a bar of rolled L iron is placed so that its bottom flange is under the soffit of the stairs; and sometimes it may be connected with the balusters when they are affixed to the outside of the stairs, after a French method that has been introduced with the object of gaining more space upon the stairs; but, in any case, it is not difficult to impart an ornamental character to the iron stiffening bar, or to the screw nuts that hold it in position. When, as in some cases, the stairs cannot be tailed in a wall at either end, it is common to pin them in between the flanges of a raking riveted girder or a rolled I joist or channel iron.

The variety of materials now used for staircases has given rise to many different methods of construction. Many modern methods of treatment have been derived from stone forms, for the oldest specimens that remain to us from antiquity are of stone. The Greeks and Romans appear to have treated the staircase purely as a utilitarian accessory to a building, and not as in any way to be regarded from an aesthetic standpoint. Among all the builders of antiquity the Assyrians and Persians best understood the imposing effect produced by vast flights of steps, as may be gathered from the remains at Nineveh and Persepolis. But the ancient modes of construction were very simple. For the most part the flights of steps were carried upon solid masses of masonry, or occasionally upon vaults, when the space underneath was to be utilized. The steps were perfectly plain, without nosings, and the modern bird's-mouth joint was conspicuous by its absence.

Much the same may be said of mediaeval staircases. In earliest forms of spiral turret staircases, a solid newel of masonry was built up in the center, and from this to the walls was thrown the vaulting, which was carried up in a spiral form, and upon which the steps were laid without being bonded either into the newel or the wall. In later examples the steps were tailed into the walls, while their smaller ends, being cut to circular form upon plan, were built one upon the other, so that they actually formed the newel.

While these few hints regarding the uses of materials other than wood for construction, are presented herewith, it is not the province of this essay to deal in other than wood in the construction of stairs. In another volume, stone, iron, concrete and terra cotta will be talked over in their relations to stair constructions.

FOURTH METHOD

In the following pages I have endeavored to show by illustrations and descriptions a variety of designs for platform stairs, so that almost any taste, or any style of building may be satisfied. I have also added some useful memoranda, which I feel assured will be welcomed by all workmen having stairs to build.

Most of the illustrations presented are from American examples, though I have thought it proper to exhibit a few of the curious or elaborate platform stairs from the Old World, not so much as specimens to follow, but simply to show to what extent of labor and ornamentation the old workmen went to satisfy their taste.

FOURTH METHOD

EXAMPLES OF PLATFORM STAIRS

It may seem lost effort to tell the workman that one of the first requisites, and the most important one, is that the carriage of a flight of stairs be built strongly and with timbers of such a dimension that any ordinary weight that may possibly be taken over the stairs will not cause the timbers or strings to "sag" or bend under the load. Often pianos are taken upstairs, and these may have a weight of anywhere from 350 to 1,000 pounds, or more, and this stair, with the weight of four or five men added who will be required to assist in getting one of these bulky instruments upstairs, will increase the weight considerably. The framework of a stairway should be made to resist a stress of not less than two tons. Strings for flights having ten or less treads, should never be less than 14 in. wide and 1½ in. thick, and these should be re-enforced by rough-cut strings 10 or 12 in. wide and 2 in. thick. One of these rough strings should be spiked or screwed to the inside of the open string, and another similarly fastened to the wall or housed string, and one or two of these strings should be placed at equal distances between the open and wall strings. The rough strings should fit accurately against both tread and riser in order to get the best results. Flights of greater length should have stouter strings and more bearing pieces. If the outside string is supported with a partition running to the floor, or the stairs have a cross partition,

FOURTH METHOD

half way in their length, then the timbers need not be so heavy; but, it is always better to err on the side of strength and rigidity than to have the frame of a stair weak and frail.

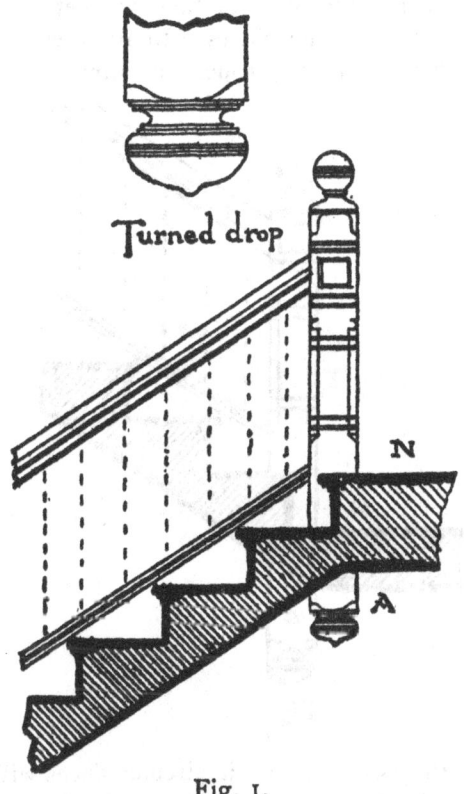

Fig. 1.

PLATFORMS

When the plan of the stair will permit, it is always better to have a platform. A platform built on posts which reach down to a solid foundation at the lower floor, or below, if necessary, is always the best. Posts may be halved at the top to receive joists or joist-bearers; or timbers may be tenoned into the posts. I prefer halving, however, as then we get the whole strength of the bearing pieces. In all cases, provision must be made for the proper fastening of the newel

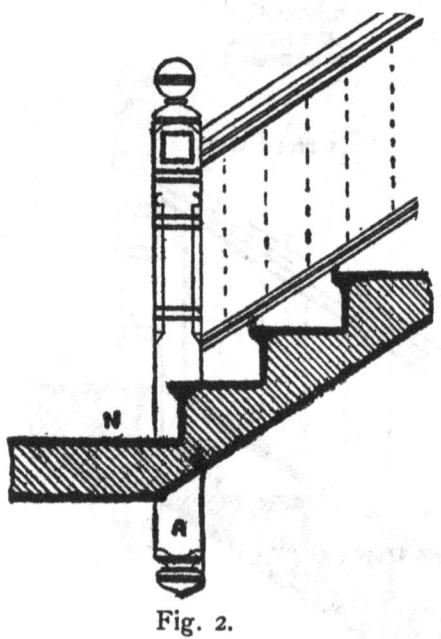

Fig. 2.

posts at the corners, and, if circumstances will admit of it, the shank of the newel post should run down below the timbering of the platform as shown at A, Fig. 1, and on larger scale above the newel. In this figure the string is shown, also the lines of balusters. There is a sub-rail in this example, which is placed just

FOURTH METHOD

above the line of nosings. A device of this kind allows a broom or brush to sweep clear through to end of step, to clean off dust without being obstructed by balusters. The platform is shown at N which may be continued to suit conditions. A lower platform which may belong to the same stairs is shown at Fig. 2. Here I show the drop of the newel A, reaching down further than the one in Fig. 1. The platform N may be extended to any length suitable to the requirements. Newels running down in the angle formed by the angle of the apron of the platform and the outside string, should be well secured to both the timber of the platform and the string. This can best be done by inserting a handrail bolt in the newel and leaving the end projecting out to pass through the timber, and another one should be placed so that it will pass through the string. Sometimes the newel is placed in position before the string is put up, and the center line of balusters is made to coincide with the center line of the newel. This is an excellent method if the stairs are open under the string, for then the "drop" can hang below the apron and string. The newel can be gained out to the proper depth over the joists, and the apron can be fitted in nicely to build against the shank of the newel post.

AN OPEN NEWEL STAIR

I show, at Fig. 3, the ground plan of an open newel stair having two landings and closed strings. The dotted lines show the carriage timbers and trimmers, also the lines of risers; while the treads are shown by complete lines. It will be noticed that the strings and trimmers of the first landing are framed into the shank of the second newel post which runs down to the floor, while the third newel drops below the apron and has a turned and carved drop. This drop hangs below both apron and string, as shown in Fig. 4. The lines of treads and risers are shown both by dotted lines and etched sections. The position of the carriage timbers

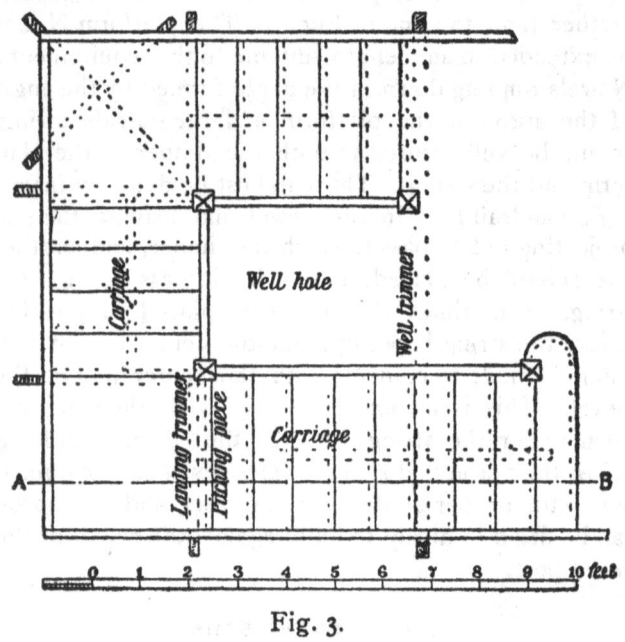

Fig. 3.

is shown both in landings and run of stairs, the projecting ends of timbers are supposed to be resting on the wall. A scale of the plan and elevation is attached to plan.

Fig. 4 shows the elevation in full with a story rod shown on the right, with the number of risers spaced

FOURTH METHOD

off. Design of newel, spandrel, framing and paneling is shown, also "raking" balusters.

Only the central carriage timbers are shown, but in

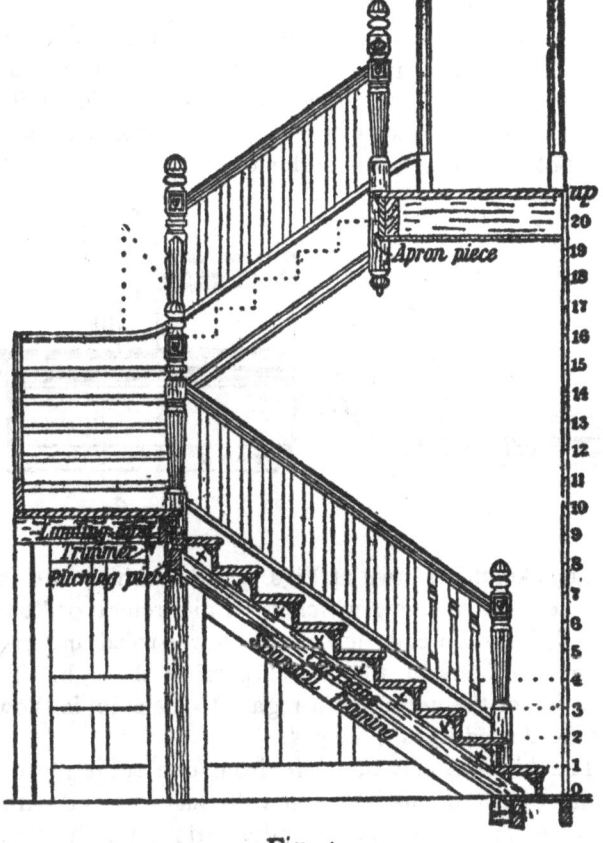

Fig. 4.

a stair of this width there ought to be two other timbers, not perhaps so heavy as the central one, yet strong enough to be of service, also to help carry the lath or paneling which may be necessary in completing the soffit. The strings being closed, the butts of the balusters must rest on a sub-rail which caps the upper edge of the outer string.

The first newel should pass through the lower floor and should be secured by bolts to a joist, as shown in the elevation, so as to insure solidity. The rail is attached to the newels in the usual manner with handrail bolts or other suitable device.

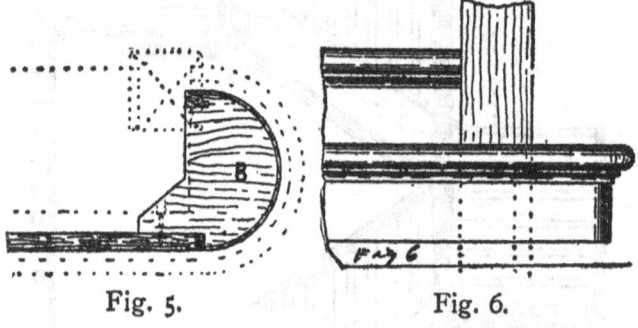

Fig. 5.   Fig. 6.

The sketches shown at Figs. 5 and 6 exhibit the end of the bottom step, which is semi-circular or "bull-nosed," also an end view of the lower tread and riser with the shank of the newel passing through. The position of the newel, with regard to the step, is shown by the dotted lines in Fig. 5.

The block B may be made from one solid piece of stuff or built up in layers and the face covered with a thin veneer, as shown in the illustration; and this finish is then the face of the riser. The nosing on the tread is worked on the end of the stuff, and the cove under the tread is worked on the end of the stuff, and the cove under the tread may be worked from the solid,

FOURTH METHOD

or it may be sprung in place if made of some elastic wood and steamed.

An examination of Fig. 4 will reveal the fact that blocks XXX are glued or otherwise fastened in the angles formed by the junction of the treads and risers. These blocks may be beveled off as shown, or they may be left simply as square blocks. This device is to give rigidity to the work. These blocks should be put in between the carriage pieces, as the latter should, when possible, fit snug to both tread and riser and go well into the angles.

Sometimes in landing stairs the rail finishes in a cap at the top of the newel; in such cases, the cap is turned, having its edge made in the same shape as the edge or moulding of the rail, as shown at Fig. 7. When this is the case, some special manipulation of the cap is necessary to have it fit properly, as shown at Fig. 8. The method of finding the proper shape of the cap is shown at Fig. 9. The

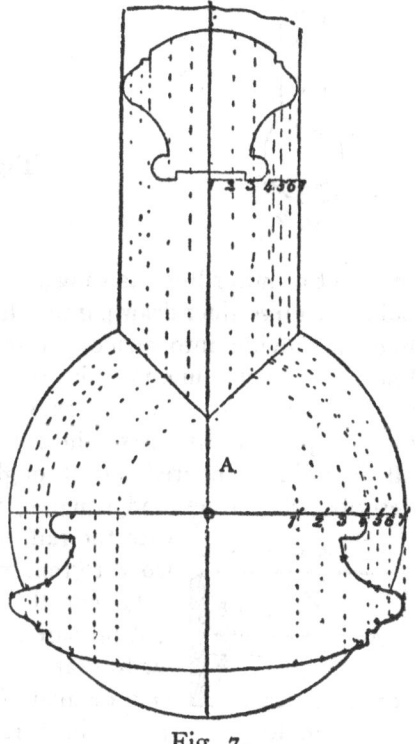

Fig. 7.

upper section shows the rail, which is supposed to be the full size; the lower section shows the cap, which may be of any reasonable diameter. Draw the plan of the cap as shown, then a section of the rail, then draw

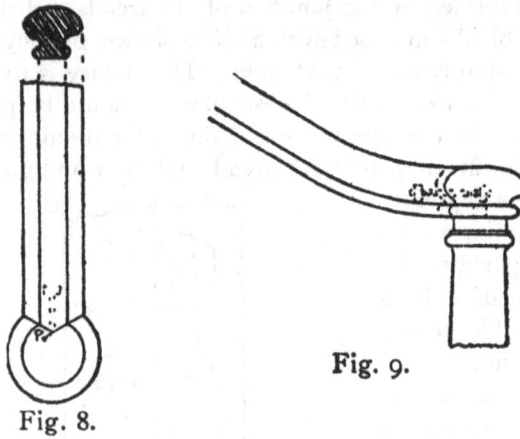

Fig. 8.

Fig. 9.

the joint or miter lines as shown, and from the outside points of these lines draw parallel lines with the central line A. Divide into spaces as shown by the dotted lines, then at the junction with the miter lines describe semi-circles as shown, until they cut the line of diameter. Square down these lines and from them prick off the points as figured, and through these prick points describe the curves and squares; then, when mitered with the rail, there will be no over wood to remove.

To cut the miter on the cap, first prepare a cutting block similar to that shown at Fig. 10, which may be made from a piece of stuff 2 or 3 in. thick and planed true

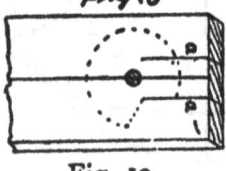

Fig. 10.

FOURTH METHOD

on the face. Gauge a center line upon it and insert a dowel that will fit snugly in a corresponding hole in the cap. Next saw two kerfs in the block parallel with the gauge line, as shown at *aa*, and at a distance from the latter equal to the square distance of the miter line *a* from the center of the cap *c*, Fig. 8. The depth of the saw cuts below the edge of the cap, which is shown by the dotted line in Fig. 10, is made equal to the length of the miter line, as shown in the plan Fig. 8. The width of the rail is marked upon the edge of the cap. The latter is then placed on the dowel and turned around until one of the marks lies against one of the saw kerfs. The saw is then run down to the bottom of the cut, and the cap turned until the other line lies on the other kerf, when the saw is again run in to meet the first cut, which finishes the miter complete. The foregoing method is the best and most economical for fitting the rail to the cap, but sometimes it is required that the joint shall be a true miter, which may necessitate some different treatment in forming the section of the cap. This is shown in Fig. 7, where full directions are given for laying out the lines for this kind of a cap

STAIR STRINGS

It is hardly necessary for me to say muc' about forming a pitch board by which stair strings are laid out, but as many of the readers of this book will be beginners in the art of stair-building, it may be well to devote a small space to this subject.

A pitch board is simply a piece of thin board, or other suitable material, and is in itself triangular. It is so cut as to represent the rise of the step and width of tread proper. The third or long side being the "run" or "going" of the stair. A sketch of one is shown in the shaded portion of the string, Fig. 11, and its application. The dotted line running through the pitch board shows the line of nosings, and the third edge or "run" of the board. The piece below this line O is a gauge or guide which is necessary to the board, for a quick laying out of treads and risers. It will be seen that the height of the riser is laid off on one edge of

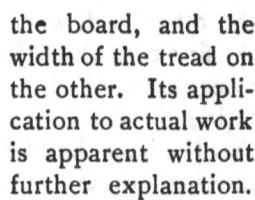

the board, and the width of the tread on the other. Its application to actual work is apparent without further explanation.

Fig. 11.

The string shown in Fig. 11 exhibits a wide tread at the bottom, a circumstance that sometimes happens—though a change of pitch should always be avoided where possible—and the string is widened out by having pieces glued to it, so that it can be "eased-off" with a gentle curve, as shown. There is also an "ease-off" near the lower floor line where the base board may butt against it. This string is, of course, a wall string, and is housed to receive ends of treads and risers. The manner of housing is shown at Fig. 12,

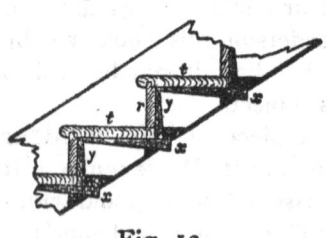

Fig. 12.

FOURTH METHOD

where the treads *t* and the risers *r* are shown in position and secured in place by means of wedges, *x*, *y*, which should be well covered with good glue before being inserted. Sometimes treads are formed with two tenons at each end which fit into mortises cut in the strings. This, however, is only applicable with closed strings.

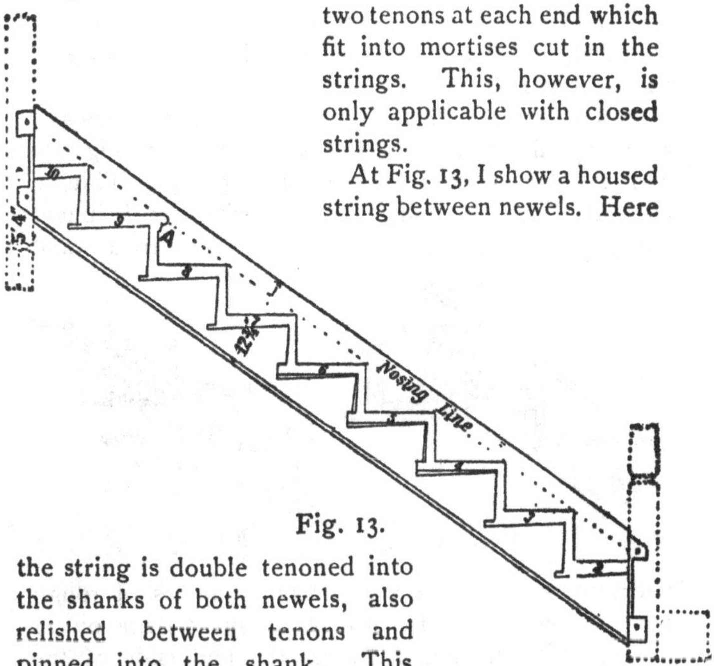

At Fig. 13, I show a housed string between newels. Here the string is double tenoned into the shanks of both newels, also relished between tenons and pinned into the shank. This string is made 12¾ in. wide, which is a very good width for a string of this kind, and the thickness should not be less than 1½ in. The upper newel is made 5' 4" long from drop to top of turned cap. These two strings are intended to be capped with a sub-rail on to which the balusters are cut or mortised in. Generally a groove the width of the square of the baluster is worked on the top of these sub-rails, and the baluster is cut to fit in this groove, then pieces of

Fig. 13.

stuff made the width of the groove, and a little thicker than the groove is deep, are cut to fit in snugly between the squares of the baluster. This makes a solid job, and the pieces between the balusters may be made of any shape on the top, either beveled, rounded or moulded, in which case much is added to the appearance of the stairs.

Two methods of arranging strings and carriages and adjusting ends against trimmers are shown at Figs. 14 and 15. The section shown at Fig. 15 exhibits a method of strengthening the stair with simple uncut strings placed against the angles of the treads and

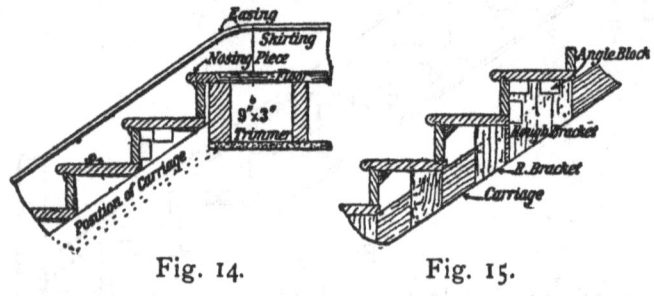

Fig. 14.   Fig. 15.

risers on the underside, and having pieces of rough boards—ends up—nailed to the rough carriage pieces and made to fit snugly against the underside of the tread and the inside of the riser. This method is not a commendable one, though much employed, as the nails may get loosened by the continual jar that a flight of stairs is subject to—a solid carriage piece is much better for the purpose.

At Fig. 16, I show an example of a cut and mitered string, with a portion of a tread, the end of which is mitered for return nosing, and dovetailed to receive ends of balusters. The other steps show how the

FOURTH METHOD

string is made and mitered to receive the riser and the tread. In the angle at the bottom tread and risers, an angular block, *a*, is shown. This tends to give firmness to the structure. The block is glued, bradded, or screwed, in place. A portion of a string, partly finished, is shown at Fig. 17. On this string I show brackets which are about $\frac{5}{8}$ of an inch thick, and

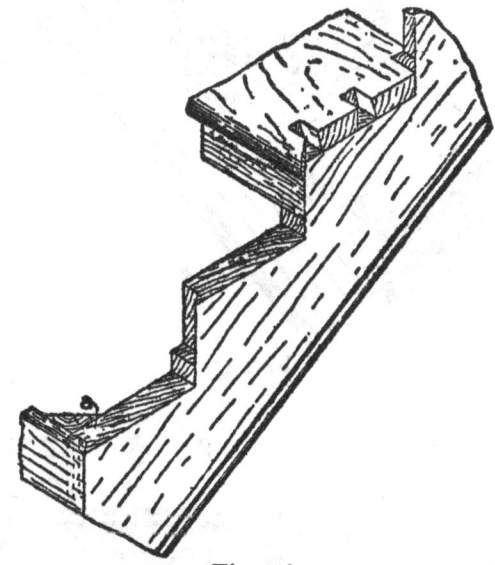

Fig. 16.

which are planted on the string. The brackets miter with the ends of the risers, and the ends of them which abut the miters should be the same length that the riser is wide, as shown at *b*. The treads must be left long enough to reach over the edge of the brackets; and the nosings and coves must also be long enough to cover the brackets as shown at *c* and *b*. The projection of the mitered riser is shown at *a*.

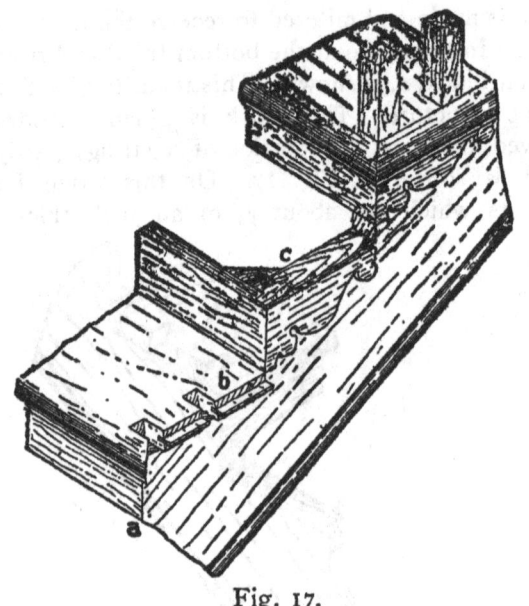

Fig. 17.

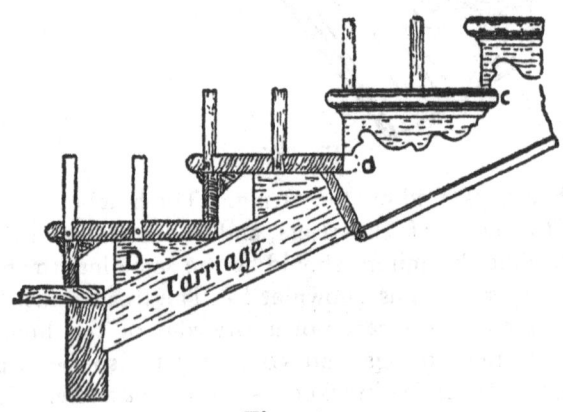

Fig. 18.

FOURTH METHOD

An end portion of a cut and mitered string, with a part of string removed, is shown at Fig. 18 in order to give an idea of the method of construction. O and C show the returned nosings, and the manner in which

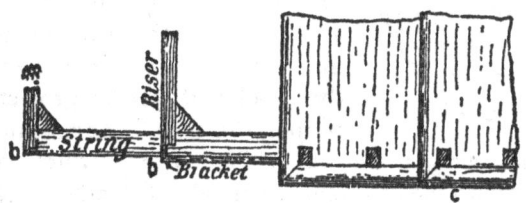

Fig. 19.

the bracket terminates on the nosing; D shows a rough bracket nailed on a rough carriage piece which is a device intended to take the place of a solid cut carriage string. The balusters are shown as being dovetailed into ends of treads.

The illustration show at Fig. 19 is simply a plan of Fig. 18, and shows the position of the string, bracket, riser and tread. The manner of mitering the riser, string and bracket is shown at *b*, and C shows the miter of the nosing at the angle of the step.

The return nosings should be fastened to the tread either by dowels or by a feather or slip tongue. The

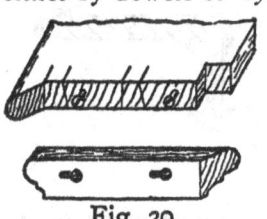

Fig. 20.

manner of doweling the nosing is shown at Fig. 20. Slot screwing may be employed for this purpose, particularly if the treads are hardwood and the work is to be polished. These screws are first screwed solid into the nosing—that end of the bolt being cut like an ordinary wood screw—and a pocket or pockets are cut in on the underside of the tread, to receive a nut, which is used to tighten up the joint when the nosing is put in place, just as the butts of handrails are fastened together. The nosing shown at Fig. 21 is fastened

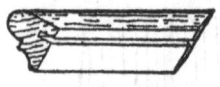

Fig. 21.

to the end of the tread by aid of a tongue or feather which is glued into a groove made in the end of the tread and left to project the proper distance. The nosing is also grooved, as shown in Fig. 21, and which corresponds with the groove in the tread; the feather is glued, after which the nosing is driven in place while the glue is warm. Many workmen put in these feathers with the grain of the wood "on end," that is, with the grain in the feather at right angles to the grain in the nosing. This, I think, makes the better job. On cheap stairs the nosing is simply nailed on, the heads of the nails "set" and the nailholes afterward puttied up.

A very good method, though rather costly, of connecting tread, riser and cove, is shown at Fig. 22,

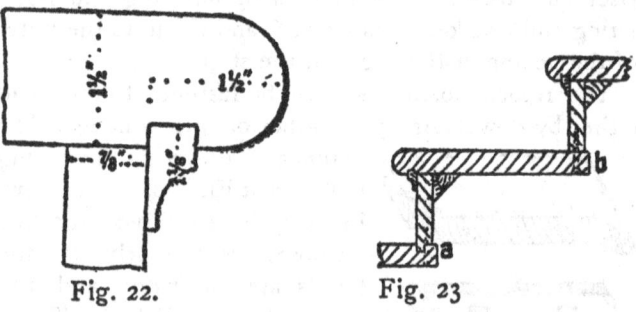

Fig. 22.   Fig. 23

where the cove is glued into a groove made in the tread. When the work is put together the cove is braded to the riser, which ties the tread down solid to

FOURTH METHOD

the riser. The completed work is shown at Fig. 23, the lower tread *a* having the riser tenoned into the tread. At *b* the tread is screwed from below to the edge of the riser. This makes strong work. The usual

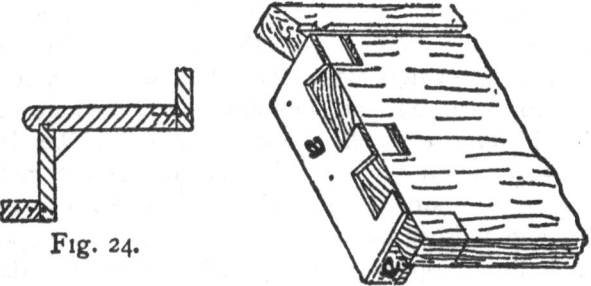

Fig. 24.

Fig. 25.

method of building a step is shown at Fig. 24. Here the riser is tongued into the tread above and runs down below the lower tread, but fits close to its edge to which it is nailed as shown in the upper portion.

At Fig. 25, I show a quick method of marking the ends of the treads for the dovetails for balusters. The

Fig. 26.

templet marked *a* is made of some thin stuff, preferably zinc or hardwood. The dovetails are marked out as shown, and the intervening spaces are cut out, showing the dovetail portions solid. The templete is then nailed or screwed to a gauge block, *e*, when the whole is ready for use. The method of using is shown in the illustration.

At Fig. 26, I show an exceedingly good method of fixing balusters before the rail is put on. A thin bar of iron, D, is spaced off and drilled, with the small hold over the center of the baluster. Screws or nails are then driven into the baluster as shown, through the holes in the iron. The rail is grooved to suit the thickness of the iron bar and laid on as shown. Holes should be drilled here and there between the balusters, and screws put through them into the wooden rail, which makes the whole work very solid. The iron bar should just be the width of the balusters, and the groove in the wooden rail should be deep enough to admit of the whole thickness of the metal.

Often the bottom tread of a stair takes a quarter turn and finishes against the base of the newel post. When this happens, some special work on the riser, tread, and nosing is sure to be required. Fig. 27 shows how the difficulty is dealth with. The riser is left the

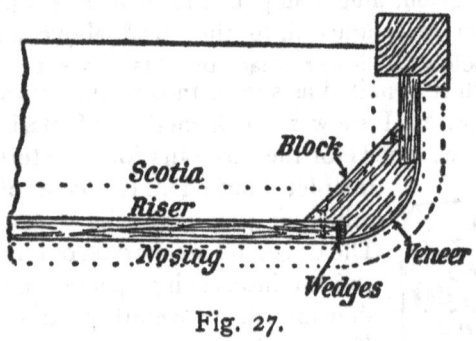

Fig. 27.

whole length of curve and return, but is cut out where the curve occurs and thinned down to a quarter of an inch. A solid block, as shown, is fastened to the floor in the proper place, being curved to the right shape. The newel is put in position, but is rebated out the depth of the riser, as shown. The thin portion of the riser is then steamed, glue is put on the block, and on

FOURTH METHOD

the thin portion of the riser that sits against the block, and when all is ready the short end of the riser is forced in the rebate shown in the newel, and is gently bent around the block as will be seen. If the work is done well, the job will be complete and satisfactory. Some workmen have concave cauls, or pieces of wood, cut out to place against the face of the work and wedge them tight against the curved riser, by any device that might suggest itself.

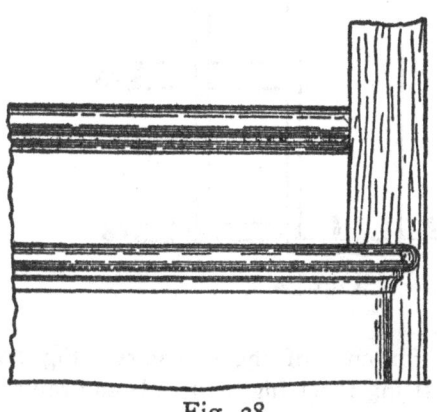

Fig. 28.

An elevation of the finished step is shown at Fig. 28, where the rounded tread and riser are seen returned against the newel post base.

The plan of a portion of stair, as shown at Fig. 29, is given at Fig. 30. Here is seen the ends of the treads as mitered, the letters WS indicate

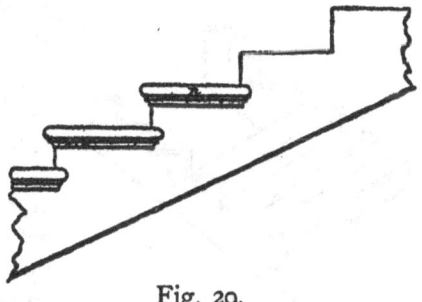

Fig. 29.

the wall string, RS the rough string, and OS the outside string. The miters of the risers are shown at *aa*, which gives the miters as being cut against the string. The square spots shown at B, B, B, B, are the

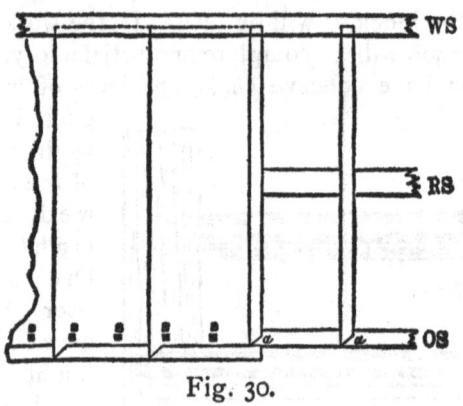

Fig. 30.

dovetail mortises for ends of the balusters. Fig. 29 simply shows the string receiving the treads and return nosings.

The illustration shown at Fig. 31 represents a portion of a flight of stairs, having cut strings, S, S, on

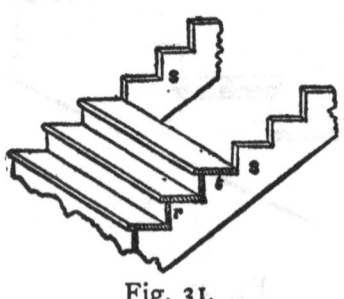

Fig. 31.

each side. The tread is shown at t and the riser at p. These are a cheap kind of stair and are nailed together. This class of stair is generally intended to fit in between walls or partitions, the strings being spiked to studding or to board timbers or wood bricks, as the case may be.

FOURTH METHOD

HOW TO DETERMINE THE RISE AND GOING OF A FLIGHT OF STAIRS

I have taken the following from Ellis' *Practical Treatise on Joiner's Work*, because it seems to me to be about the best thing written on the subject, at least, the best I have come across. "The amount of going and rise given depends chiefly upon the amount of floor space allotted to them, and upon the height of the story; but subject to these restrictions, there is room for considerable variation. To obtain a stair that shall not be fatiguing or awkward to ascend or descend, the going should bear a certain ratio to the rise. Various methods have been proposed by writers on the subject to obtain the ratio, of which the following are the best known and most practiced:

"1. It is assumed that the average length of step in walking on the level is 24 in., and that it is twice as difficult or fatiguing to climb upward as it is to walk forward. From these premises it is deduced that one going one step forward, plus two rises or steps upward, should equal 24 in., which put in the form of a rule becomes,

"To Find the Rise When the Going Is Known.—Subtract the given going from 24 in., and divide the remainder by 2 for the rise.

"To Find the Going When the Rise Is Known.—Multiply the given rise by 2, and subtract the product from 24. The remainder is the proportionate going required.

"2. The product of the going and rise multiplied together is to equal 66. Example: Going 11 in. x 6 in. = 66, and 7 in. rise x 9¾ in. = 66. Rule by this method: Divide 66 by the given rise or going to ascertain the proportionate going or rise.

"3. Assume 12 in. going and 5½ in. rise as a standard ratio. To find any other, for each addition of ½ in. to the rise, subtract 1 in. from the going. Example: Rise 6 in., going 11 in.; rise 7 in., going 9 in. It will be noted that by this method the sum of 2 rises plus the going equals 23, which affords an easier stair than the first-mentioned method.

"When the total rise of the stair is known, as shown by the story rod, Fig. 32, and the approximate rise of the step is given, the exact rise is obtained by calculation, thus: Reduce the total height to inches, and divide it by the desired rise. If there is no remainder, the divisor will be exact rise, and the quotient will be the number of risers required. If there is a remainder, again divide the sum by the quotient, discarding the fraction, and the result will be the exact rise. For instance, let the height of the story be 10 ft. 6 in., and the proposed riser 6½ in. 10 ft. 6 in. = 126 in. ÷ 6½ in. = 19 with 5 remainder; then 126 in. ÷ 19 = 6⅝ in. full as the rise, and the proper ratio of going to this, as found by the first method, is 6⅝×2=13¼−24=10¾; but the exact going is found by dividing the plan into 18 equal parts, as there is always one less tread than the number of risers, in consequence of the landing acting as tread

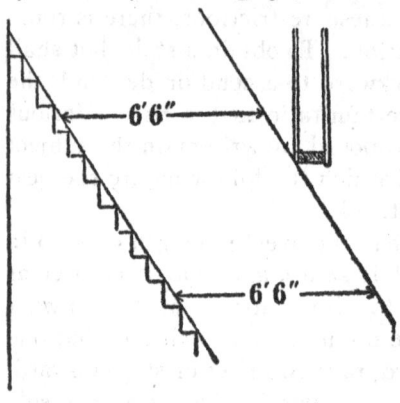

Fig. 32.

FOURTH METHOD

for the last riser. No arbitrary rule can be given for the treatment of the plan, which must be subject to circumstances. Every attempt should be made, however, to dispense with winders, which should be introduced in case of necessity, when they are better placed at the top of a flight than at the bottom."

All stairs should be so devised that not less than 6 ft. 6 in., head-room between tread and trimmer, is given, but, as shown in Fig. 32, it is much better to give this much space from the going line to the trimmer, then

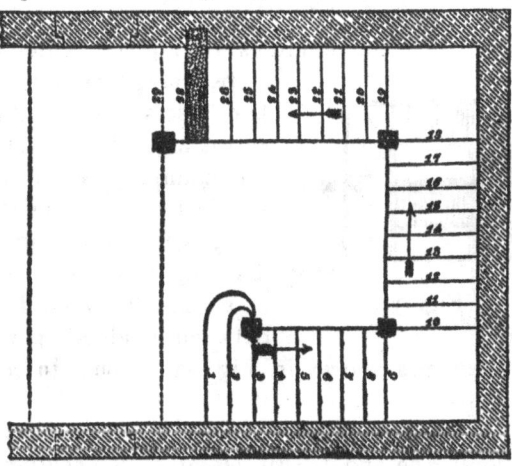

Fig. 33.

there will be no danger of a tall man striking the trimmer with his hat on his head. There will be cases, of course, where to give so much space for head-room will be impossible, but in ordinary stairways any less space than that determined will surely prove unsatisfactory.

VARIOUS PLANS FOR STAIRS

A newel or landing stair can be devised that it will serve the purpose for almost any possible contingency;

and in order to make this plain I show a number of plans, which I am sure will prove of use to the general workman as well as to the stair-builder, as they offer hints and suggestions for dealing with almost every condition and situation that are likely to present themselves in preparing plans for stair runs which are intended to be of the platform style.

In Fig. 33, I showed a plan of a stair having two landings, and a circular-ended step, with dotted lines showing trimmer timbers. At Fig. 34, I show another plan with the order of going reversed, and with the flight between the landings having a less number of steps. Fig. 34 only shows five risers, while Fig. 33 shows nine risers. The latter example also shows the two lower steps rounded off to fit against the newel post. The flight shown in Fig. 34 is supposed to be built in between studded partitions while the stairs shown in Fig. 33 are built in between brick walls.

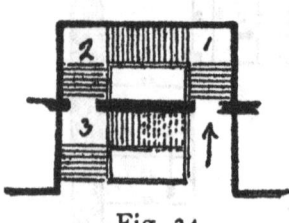

Fig. 34.

Fig. 34 shows a plan of stair in the Time St. Depot, Liverpool, England. This is rather a peculiar stairway, as from the third landing the stair starts off in two directions so as to reach different parts of the building more conveniently. The plan shown at Fig. 35 illustrates an elaborate entrance and stairway to the National Gallery of Arts, London. This is a peculiar stairway inasmuch as there are two flights leading up to a large platform where the upper flights broaden out and carry the visitor to the upper floor either to the right or to the left. This is rather an

FOURTH METHOD

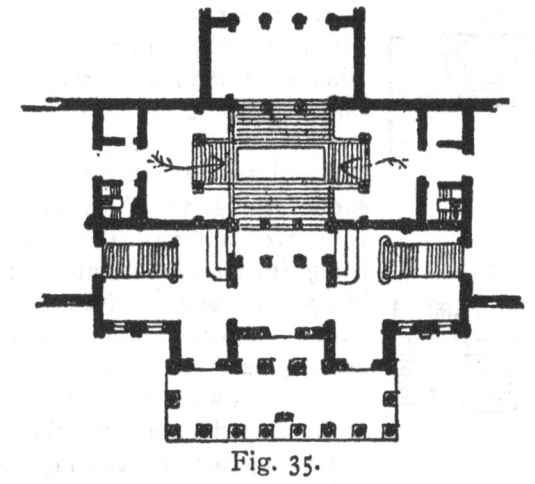

Fig. 35.

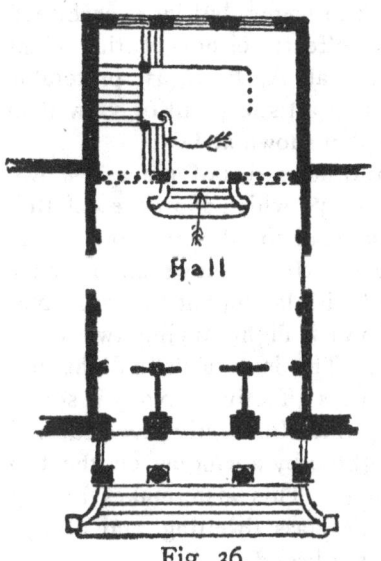

Fig. 36.

ingenious arrangement and might be made use of in many instances for public buildings.

An effective arrangement for a hall stair is shown at Fig. 36, where a short flight of stairs lead to a raised dais from which a second flight of stair springs, in which there are two landings. The rail over the lower flight runs from two starting newels, and fin-

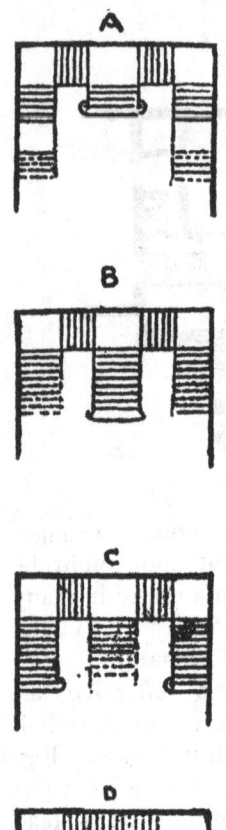

Fig. 37.

ishes against columns having their base on the plinth of the dais. The windows in the rear of the hall are filled with art glass, and the whole is artistic and impressive.

A series of sketches for plans is shown at Fig. 37. A shows a stair with five landings, the first step being situated in the center. B exhibits a similar stair with three landings. C shows a stair with three landings and two starting flights leading to a wider flight above. D shows a flight with two landings and having but one starting point. This is a common kind of stair and much in vogue, but lacks architectural effect; either of the plans shown at A, B, C, is preferable from an artistic point of view than the plan shown at D.

Another series of plans is shown at Fig. 38, which show the relation of the stair to other portions of the house. No. 1 shows an ordinary flight with landing at the top. No. 2 shows a flight having two landings. This is an artistic flight and is always effective. No. 3 is something like No. 2, only reversed, and is lighted by a window on the top landing. This also, makes a very effective stair for a middle class dwelling, and always looks well if finished in hardwood.

FOURTH METHOD

Another series of plans is shown at Fig. 39, with parts of the plans of the buildings along with them. No. 1 shows a stair with two landings and a "step-off," on the second landing at O, leading to rooms over the kitchen which are used for the domestics. No. 2 shows a very different arrangement, the stairs being built in an inner hall which leads into a conservatory. The plan shown at No. 3 is very much in vogue at the present time, and is really a very good style of stair.

A very good "lay-out" for a hall and stairway is shown at Fig. 40. Entrance to dining-room, drawing-

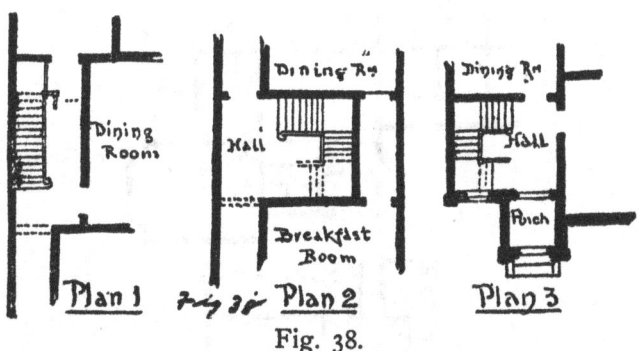

Fig. 38.

room and library is gained direct from the hall, and the hall is entered from the street by way of vestibule as shown. Access also to kitchen and outer offices, is also obtained from the hall. The stairs are well arranged with wide platform and is well lighted by two windows over the platform; the windows being filled with suitable art glass. This particular arrangement of hall, stairs and rooms is worthy of being thought over by those of my readers who may have anything to do with designing floor plans.

I think I have now given a sufficient number of plans

to enable the workman to "lay-off" a stairway that will "fit" in almost any situation, or at least to suggest to

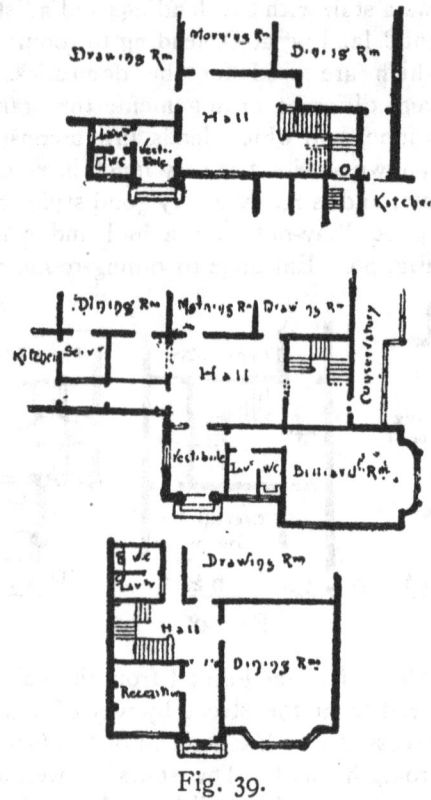

Fig. 39.

him how the difficulty may be worked out, so I will now leave this subject, feeling that I have done it full justice.

NEWELS, NEWEL POSTS, BALUSTERS, AND ORNAMENTAL BALUSTERS

The different styles of newels and newel posts are without number, and I will not make any attempt to describe or illustrate more than will give the workman an idea of those most common in use at the present time, and a few elaborate ones now in existence that were designed and set up by old workmen.

The sketch shown at Fig. 41

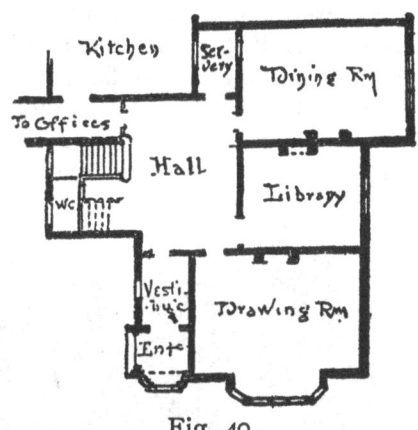

Fig. 40.

Fig. 41.

136   COMMON-SENSE HANDRAILING

is a design for a large hall and stairway, and is in Mercer's Hall, London; the stair is supported on columns, and shows three landings with balusters and newels. It will be noticed the long flight is last, leading up to the floor. This seems to be a rule with English stairways, as it is argued that there is a longer

Fig. 42.

rest at the top, therefore the long rest comes after the long rise.

Another and still more elaborate staircase is shown at Fig. 42. This is a stately and palatial class of stairs and consists of central flights branching off into lateral flights, surrounded by a gallery separated by columns or arches. This shows the main stairway and hall of the opera house, Paris, France.

FOURTH METHOD

The two lateral flights lead to a spacious landing, from which a wide curvilinear-shaped flight of stairs ascends with wing stairs to the gallery. The elegant and graceful lines of this staircase make it almost unique among great modern examples. The architect has introduced the ramping arch below the flights, and by curving the balustrades outward has given ease of ascent and grace of outline. Round the gallery rise coupled columns of red polished granite with Ionic capitals carrying entablatures and arches, above which runs a rich truss cornice. Over the cornice on each side are rows of lunettes, surmounted by the fine vaulted quadrangular domical ceiling. Much of the grandeur of this staircase is due to the surrounding gallery, which impresses the visitor on ascending. The magnificence of the *escalier d'honneur* is heightened by the arrangement of the minor stair and the open loggia and vestibule. As a model of planning the Paris Opera House stands pre-eminent. It forms a long rectangle, flanked by projecting annexes, which give much variety to its length. There are three parts or divisions symmetrically disposed to the major and minor axes: the stage occupying the whole breadth of the building; the theater proper, or auditorium, forming the center of the building, and including the grand staircase; and, lastly, the promenade and open loggia in front. The staircase hall forms a square and complete structure between the foyer or promenade and auditorium, and is surrounded by corridors. The plan of this building is an instance of the centralizing mind of the French; every organic function is expressed in the structure. In a large public building the staircase performs an essentially distinct and public function, and too much prominence cannot be bestowed upon it.

138 COMMON-SENSE HANDRAILING

The newels and panel-carved balustrade shown at Fig. 43 exhibits an extremely rich example of sixteenth century work. All this is taken from work still standing in a house at Greenwich, England.

Fig. 43.

The interior of the house is very nearly in its original state. There is a very curious internal court. The rooms have several good door-cases and ornamental plaster ceilings.

Fig. 44 is a plan of the stairs, showing the double

FOURTH METHOD

approach. The sketch, Fig. 45, shows the half section of handrail. Fig. 46 shows section of the entablature under steps with the carved pendant under newel; and Fig. 47 shows the top of one of the newels at large.

Stairs finished in this manner are coming into vogue again, and a number of similar ones have been built in our larger cities where cost was only a secondary consideration. Stairs of this kind, to be effective, should be massive in appearance; the newels should be heavy and the carving done in the solid. The newels in the

Fig. 44.   Fig. 45.

example shown, are rather light in appearance, but the whole mass is quite impressive. The details, Figs. 45, 46 and 47, are left to a larger scale than the main illustration so that the workman may the easier enlarge and copy them for actual work, if he so desires.

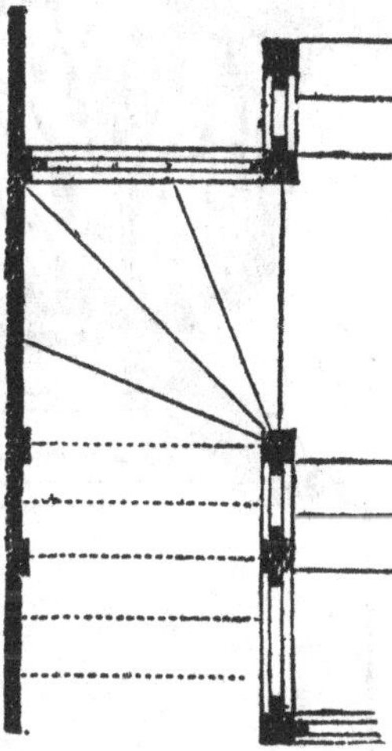

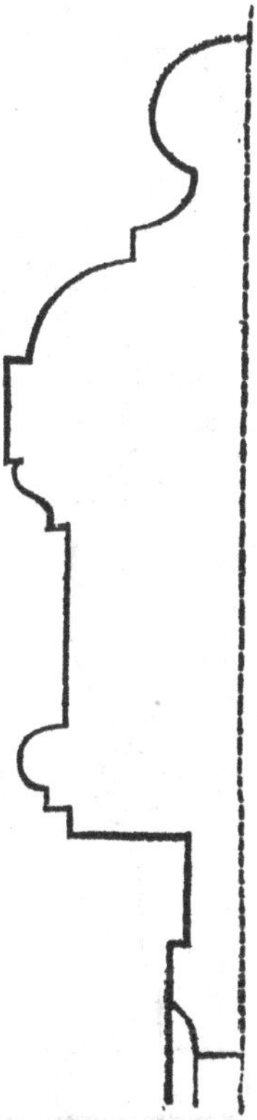

140 COMMON-SENSE HANDRAILING

Fig. 46.

Fig. 47.

FOURTH METHOD

A flight of stairs with newels, having carved balustrade, strings and newels, is shown at Fig. 48. This is a

Fig. 48.

French design, and has a very broad rail elaborately wrought. The carved string is a special feature of this example, so also are the square carved vases that surmount the newels. It will be noticed that the newels in this example are much larger in section than those shown in Fig. 43. This is, I think, a gain in appearance. It will be noticed that the central ornament on the finished faces of the newels is partly turned and partly carved, so must necessarily be planted on, as are perhaps the other carvings.

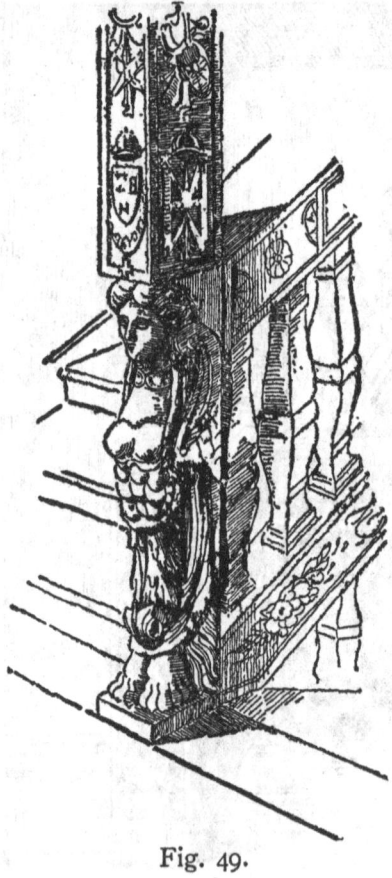

Fig. 49.

A curious example of a newel post is shown at Fig. 49, which represents a portion of a stairway in the Hotel Cluny, Paris, France. The stair has a close heavy rectangular string with carved rosettes sunk flush on both sides. The balusters are square in section and are massive, with all the mouldings worked on the rake of the stairs on two sides, and square across on the upper and lower sides. This style of baluster is quite common in Europe and

FOURTH METHOD

is really quite effective. Another peculiarity of this

Fig. 50.

stair is the upper portion of the newel which runs up to the ceiling, and is carved on its four faces with various emblematical devices. This gives the whole work rather an odd appearance.

A modern stairway with landings is shown at Fig. 50. This is in Colonial style and has a very cozy and inviting appearance. This example is taken from a New England house, and is noted for the width of stairway and breadth of tread; the rise being little more than five inches. A peculiarity of this stairway is the twin newels at the main landing. This is a departure from the general practice, and is employed here for architectural effect. The newels are plain, yet they are quite effective. The balusters are heavy and placed pretty close together.

Another staircase of recent design and intended for a house in Philadelphia, Pa., is illustrated at Fig. 51. This is in Colonial style and shows a semi-circular finish on end of bottom tread and riser. Sections of the newels are octagon, and the rails are finished against the newels with a "goose-neck" curve and square. The landing turns at right angles.

The illustration offers a number of excellent suggestions for work other than for the stairs.

The examples shown of stairs in place, I think, are quite enough to give the workman an idea as to their treatment, so I will now offer a few designs for newels and balusters, and a few remarks as to their treatment.

There is no end to designs for newel posts, yet it is a strange fact that when a workman undertakes to design a newel for any particular stair he may be building, he finds it very difficult to decide upon the exact design he has in mind. This is owing to the fact that the workman possesses a certain amount of art instinct, and his mind requires for its satisfaction a newel suited to the fitness of the situation. An

FOURTH METHOD 145

Fig. 51.

146   COMMON-SENSE HANDRAILING

elaborate stairway demands an elaborate newel and baluster, yet both must be in keeping with the surroundings as well as in keeping with the stairway. A good

Fig. 52.

illustration of this is exemplified in Fig. 51, where all the work seems to have a like character. Another illustration of the true fitness of things is shown in

FOURTH METHOD

Fig. 52. In this example there is a quiet Quaker-like repose both in stairway and finish, yet the observer cannot but be impressed with the wealth and dignity of the whole design. The newels are comparatively plain, yet they are effective and seem to be in the proper place. The whole stairway, paneling, strings, newels and balusters are in same finish as the woodwork in the hall. The mantel, which is quite plain, is chaste and in keeping with the general design. The first step is semi-circular at each end being returned against the newels. This stair and its appointments are well fitted for a Colonial house of the earlier period; in fact, it would do for almost any period of Colonial architecture.

The sketches shown at Figs. 53 and 54 exhibit styles of stairs, newels and balusters in Elizabethan style. The balusters in Fig. 53 are square, and those running down the strings are worked on a rake; all the members being cut on the same incline as the "lay" of the string and rail. This is, of course, expensive, but it gives a fine appearance to the stairs. The sketch shown in Fig. 54 is taken from

Fig. 53.

French work. It shows three balusters on each tread and carved brackets under end of nosing of treads. The plan of the rail is shown in the sketch at 3. The balusters are turned spiral, as will be seen. An alternate baluster is shown at No. 2.

I have given the proportion of riser and treads in an earlier paragraph, and it may be well at this point to say something regarding the height a rail should be from the tread. An authority says "that the height from the treads at the nosings to the upper part of the handrail should be 2 ft. 7½ in.; at the landings the height of half the riser should be added, this variation in the height conducing to ease and safety, a person requiring more protection when he is standing on a landing than when ascending a stairs. Two balusters are generally placed on every tread, one on the same plane as the riser. In the old close string staircases, where massive rectangular or turned balusters are seen, one to each step is common. Of handrails, the moulded is the handsomest; a roll member with cymas on each side, and a deep rail moulded at the sides with ovolos or astragals, is commonly met

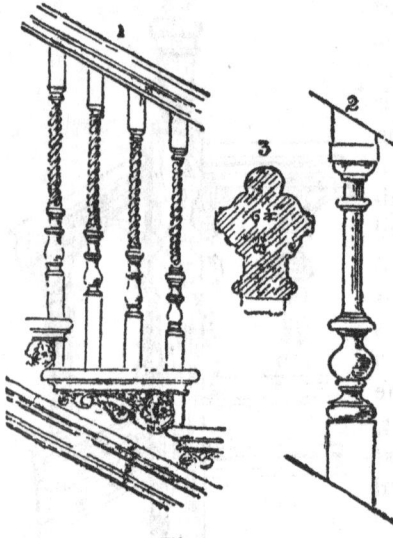

Fig. 54.

FOURTH METHOD

with in the older examples and is very effective. (See sketch 54, 2, above.)"

In a recent text book on building construction the student is instructed, before planning a staircase, to know the position of doors and windows surrounding, so that the steps and the first and last riser may be

Fig. 55.

fixed accordingly. Advice of this kind is very well when a staircase has to be fitted in a given space; but the architect, in planning and designing the stairs, ought to proceed quite differently. He should first plan his stairs, as being the most important thing, and then arrange the hall thereto. The "going" of the flight or the positions of the first and last risers should not be made to depend on the doorways and approaches, but these should be adjusted to the risers. Given a space to design a stairs in, it may be a mode of proceeding in some cases; but if any attempt is made to give the staircase a character of its own, its design should be undertaken *pari passu* with the hall in which it is to be placed. No architectural arrangement can

Fig. 56.

be possible under any other conditions.

Another style of stair is shown at Fig. 55, having turned newels at the bottom and square ones at the landings. This is a purely Colonial stair with the conventional shaped newel and baluster. All the rails in stairs of this kind are made straight and are fastened into the newels with either tenon or stair bolts or both, and glued.

FOURTH METHOD

The stair shown at Fig. 56 is taken from an English example of the Georgeon period. Both rail and newel are heavy, the latter being surmounted by a carved finial. In this example the risers are low and the treads wide, a characteristic of nearly all English stairs, a custom well worthy of imitation. The heavy newels employed in this stair give the whole design a

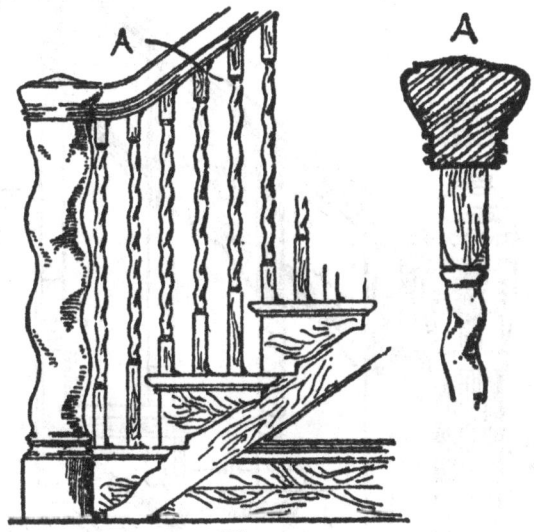

Fig. 57.

massive and substantial appearance. Of course, where a stair of this kind is intended to be placed, it must have plenty of room, as the run or "going" will require a good stretch owing to great width of tread, and the hall or reception room must be large to accommodate the stairs and be in keeping with them.

At Fig. 57, I show a portion of a stair having serpentine newel and baluster. This style of work is very

152 COMMON-SENSE HANDRAILING

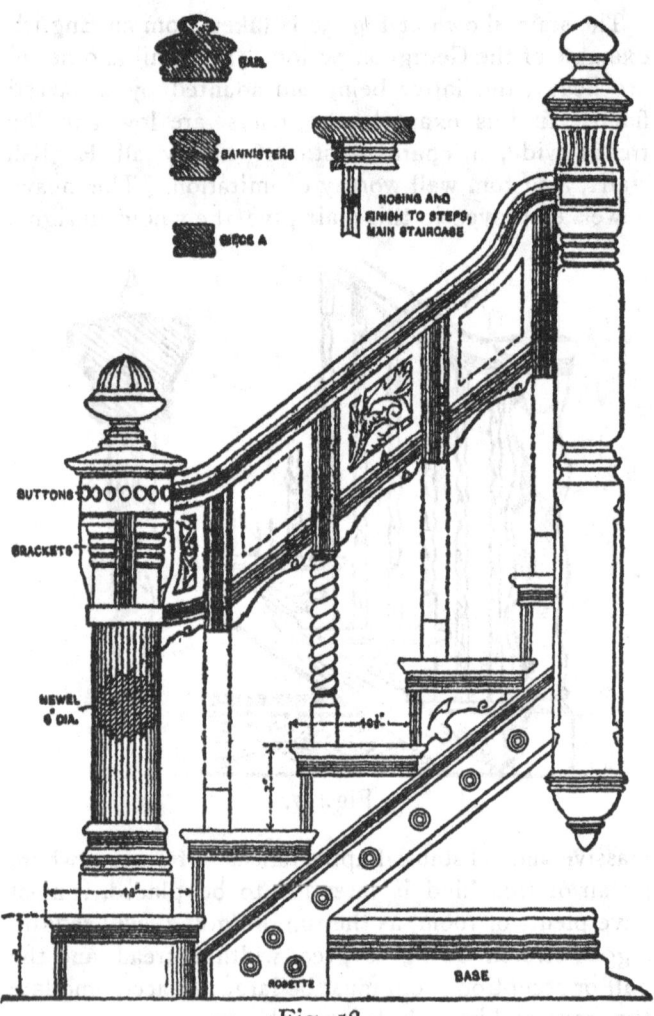

Fig. 58.

troublesome and is not much in favor, as the results are not in proportion to labor expended. A, A shows

FOURTH METHOD 153

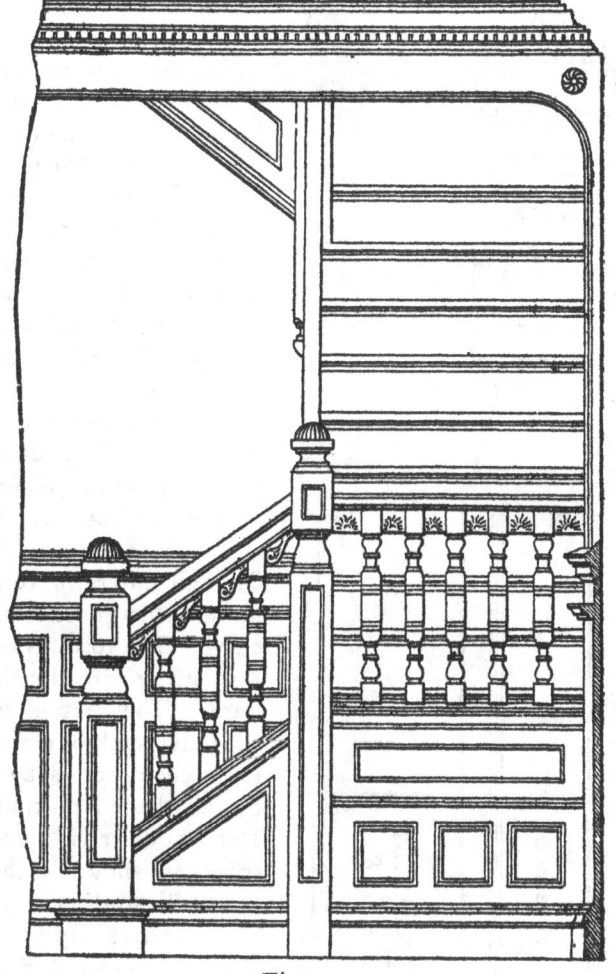

Fig. 59.

the style of rail which generally accompanies this style of ornamentation.

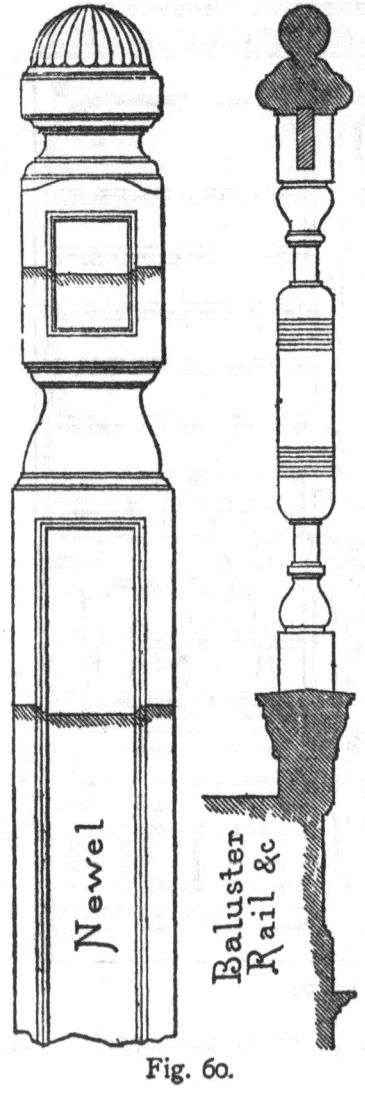

Fig. 60.

The illustration which is shown at Fig. 58 is adapted from *Carpentry and Building*, and is a good example of a modern stair. The paneling between the baluster at the top, marked A, is perforated. The treatment of the string is somewhat unusual, and it will be noticed that the nosings on the treads are worked to a flat ogee. The drop newel is quite plain, except the top, which is very nicely wrought. The rail enters the top newel with a goose-neck curve. The rosettes on the string are let in flush. The section of the fluted shaft of newel is circular, as shown by the shaded portion. Details of rail and treads are shown on the top of illustration.

Another style of stairs is shown at Fig. 59. A part of the paneled wainscot is shown, also lower spandril and

FOURTH METHOD

Fig. 61.

paneling of platform. Fig. 60 shows a portion of the newel and a baluster with section of rail drawn to a larger scale.

Fig. 62.　　　　　　　　Fig. 63.

Fig. 61 shows a built-up newel, and a couple of tread ends and a part of baluster; it also shows the rail with

FOURTH METHOD 157

ramp entering the newel post. The bottom tread is partly returned against the base of newel.

The example shown in Fig. 62 may be put down as one seldom required in this country, though I have

Fig. 64.

seen it, or one very similar, employed on a stairway leading to a gallery or speaker's platform. It is almost a solid balustrade.

Fig. 63 is of a style often employed in and about public buildings in England, Belgium and France.

158 COMMON-SENSE HANDRAILING

In styles of this kind there is no regularity; the newels and balusters may be of a different pattern on each flight of stairs; they offer an abundance of opportunity for a display of originality of design on the part of the architect.

The example shown at Fig. 64 is taken from a stairway in St. Jacob's Church, Bruges. The newel is a carved figure which is said to be one of the finest pieces of carving in Europe. The rail and sub-rail are heavy, and the spaces between them are filled with fine carvings instead of balusters. The string is also carved with a running wreath. The whole is made of heavy oak. The work is over two hundred years old and is in excellent preservation at this date.

Fig. 65.   Fig. 66.

A couple of commonplace newels are illustrated at Figs. 65 and 66. The first is simply a turned post with an octagon base and flat facets, or neck, and surbase. The second example belongs to the so-called Queen Anne style It is neither more nor less than a square post with a few ornaments worked on two sides on a

FOURTH METHOD

rake with the line of rail, and has chamfered corners. The ornaments are worked square across the lower and upper faces from the lines of the raking ornaments where they cut the angles of the post.

The example of newel shown in Fig. 67 is from the Cincinnati school of design, of which Benn Pitman was principal. This newel was carved by a young lady, Miss Louise Nourse, and is worked to over two inches relief projecting one inch over the border. The entire height of a newel is 4 ft. 9 in. It is illustrated here as an example of what may be done by the ordinary workman if he only apply himself to the task. Newel posts offer splendid opportunities to the carver.

The newels shown at Figs. 68 and 69 are octagon in section and are rather elaborate in finish. This style of newel is often made use of, but I confess I do not like them; they seem more like

Fig. 67.

pedestals than newels, and are certainly vulgar when made up with different colored woods. They are also unnecessarily costly, as they entail considerable labor in the making up; particularly is this true of Fig. 68,

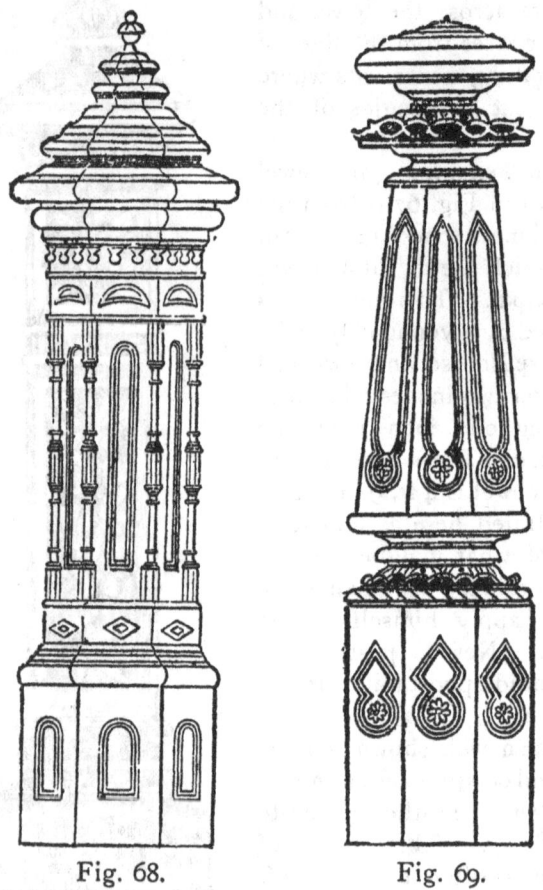

Fig. 68.   Fig. 69.

as all the mouldings must be mitered around the cap and the base. The result is not worth the labor, as the architectural effect is disappointing.

FOURTH METHOD

I will close my remarks on newels and newel posts by offering a few examples of quaint design culled from domestic and foreign sources; the example shown at Fig. 70 is taken from a stairway in Boston. This is a handsome design, but has one fault: the central column looks too much like a screw. It gives one the impression of a jack screw for raising great weights. If this column was fluted, the effect would be much more pleasing.

The carved newel shown at Fig. 71 is drawn from an example at

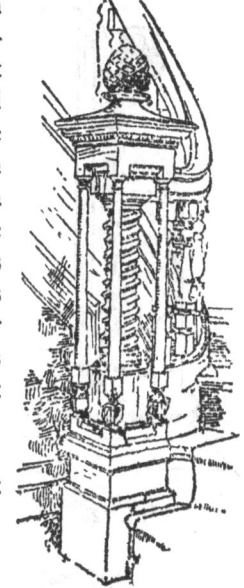

Fig. 70.

Fig. 71.

Argeles on the Spanish frontier near the Pyrenees. The one shown at Fig. 72 is at Tuz, a small town near Argeles.

The three examples shown at Figs. 73, 74 and 75 are from the same neighborhood as are those shown in Figs. 71 and 72. They are quaint and odd, and are generally placed in small narrow halls dimly lighted, and are apt to

162 COMMON-SENSE HANDRAILING

startle a stranger when he first enters. The examples offered are among the best, but there are some that rise above the head, and are topped off with hideous faces or grinning skulls and other uncanny things.

BALUSTERS OF VARIOUS KINDS

Before giving any designs for balusters, it may be well to say something about their arrangement with regard to their relation of length, rail and tread. Sometimes the architect who designs the stair may have very decided ideas as to the manner of arrang-

Fig. 72.

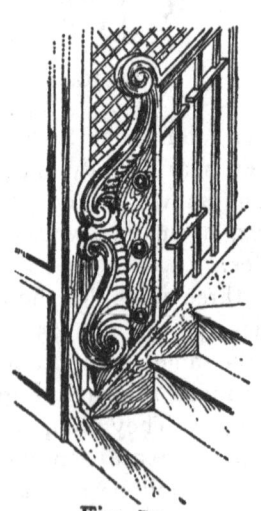

Fig. 73.

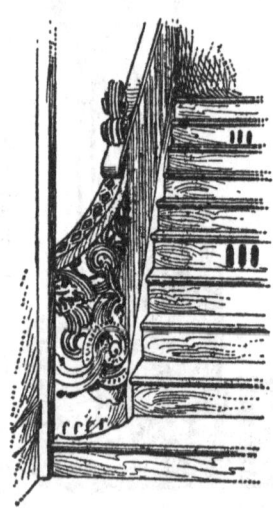

Fig. 74.

FOURTH METHOD

ing the balusters, and I give a few examples arranged differently in a stair having rail, string and baluster about the same. Fig. 76 shows one of the ordinary methods where the turnings are all of one length, and thus all the squares run parallel with the handrail. In Fig. 77 the turnings are of two different lengths, the upper squares being all of one length and running parallel with the handrail, the bottom squares being all the same length and thus each pair being parallel with their respective treads, the middle member of the turning usually being arranged as shown. A method that is perhaps not much in general use is shown in Fig. 78, where

Fig. 75.

the turnings are all of the same length and the bottom squares equal, but the bottom ledges of the upper squares of each baluster run parallel with their respective treads, producing long and short upper squares alternately, as shown. It will be seen that

164   COMMON-SENSE HANDRAILING

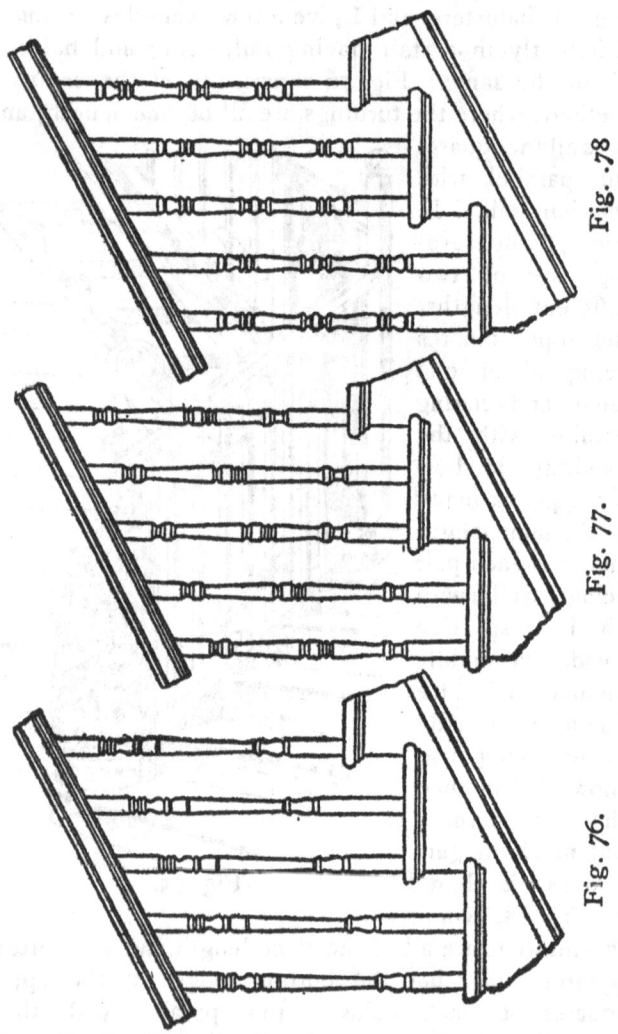

Fig. 76. Fig. 77. Fig. 78.

after all the difference in these examples is altogether in the lengths of the turned part of the baluster.

FOURTH METHOD

The patterns for balusters shown at Fig. 79 may be suggestive. Balusters of this kind may be obtained at any well-equipped factory any length or size that may be required.

A few Colonial balusters and a spiral newel are shown at Fig. 80. This makes a handsome termination for a stairway.

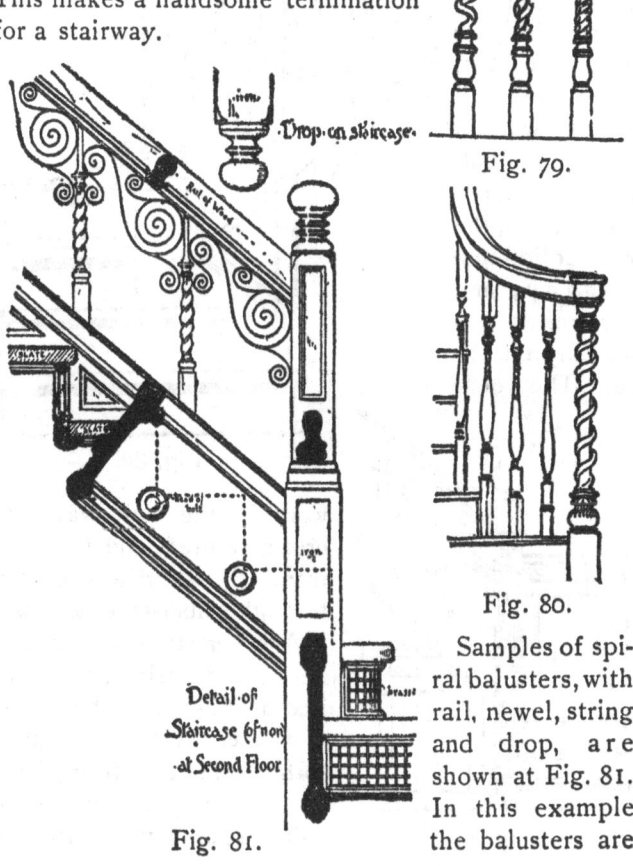

Fig. 79.

Fig. 80.

Detail of Staircase of non at Second Floor

Fig. 81.

Samples of spiral balusters, with rail, newel, string and drop, are shown at Fig. 81. In this example the balusters are shown reinforced by bent iron scroll work; this has a charming effect in many cases, and I know of one instance, in New York City, where the scroll work was of brass, the balusters enamelled cream, the rail solid mahogany, and the result was actually beautiful. The wood-

Fig. 82.

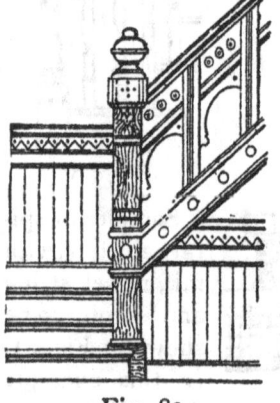

Fig. 83.

work in the hall was also cream-colored, and the light from the outside passed through amber-colored glass.

Another style of baluster, newel and string is exhibited at Fig. 82. The newel is formed at the first platform, there being three risers up to the platform. The balusters are simple, and the

FOURTH METHOD

whole illustration is given here more to show the method of raising the newel and balusters than for any other purpose.

Another style of baluster, string and rail is shown at Fig. 83. The baluster in this case is simply a square with two of its sides bevelled and cut in between the

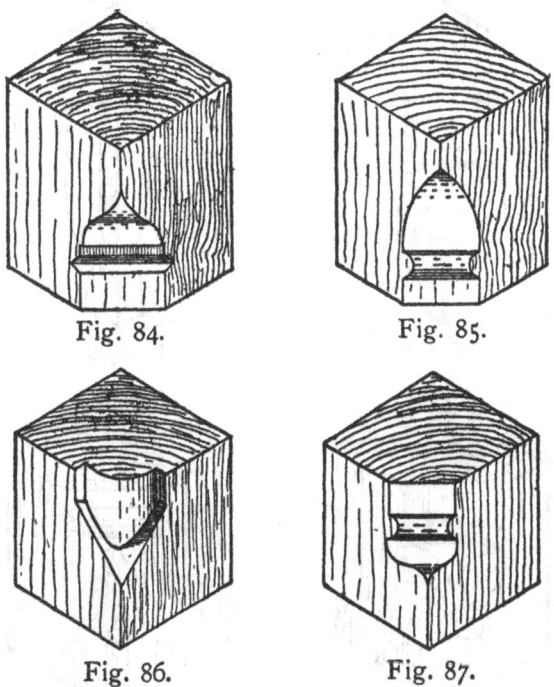

Fig. 84.   Fig. 85.

Fig. 86.   Fig. 87.

rail and the sub-rail. The little panels running raking with the rails are also cut in, or let into grooves in baluster and rail. Other portions of the illustration are self-explanatory.

These examples of newels and balusters, I think, are sufficient, as trade catalogues from factory and shop,

168 COMMON-SENSE HANDRAILING

contaning hundreds of set designs, may be obtained for the asking.

MISCELLANEOUS ITEMS

Under this head I purpose showing a few things not generally included in works of this kind, but which

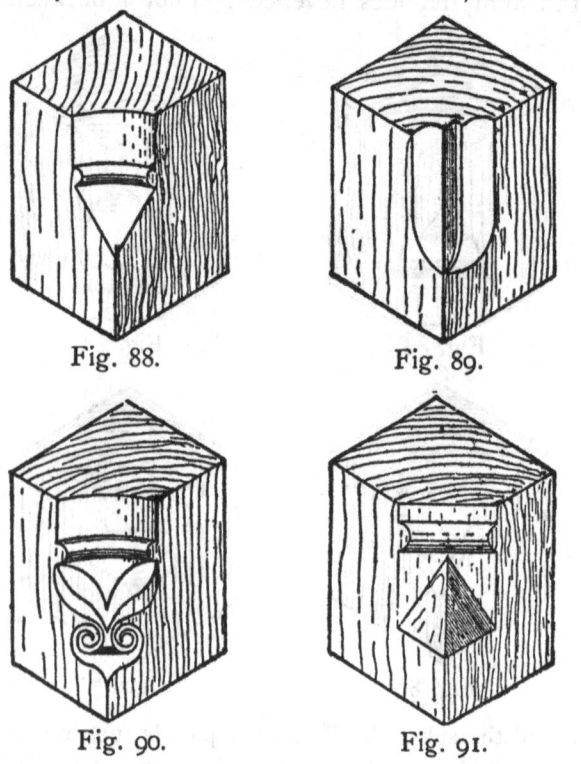

Fig. 88.   Fig. 89.

Fig. 90.   Fig. 91.

will be found very useful to the general workman as well as to the specialist in stair-building.

The illustrations shown in Figs. 84 to 91, inclusive, exhibit a number of different designs for stop cham-

FOURTH METHOD

fering. These will be found useful in determining the style of step for chamfering the corners of a newel post, and in many other instances as well. Some of these chamfers and stops are quite elaborate and will require considerable labor to work them out in good form; particularly is this true of Figs. 86 and 89, as one has a concave and the other a convex surface, and Fig. 89 has an ornamental termination.

Besides these styles of stops there are many others, the simplest of which is just a bevel ending of any pitch and the ogee ending, and several others of which nearly every workman is familiar.

The illustration shown at Fig. 92 gives the method of obtaining a reduced pattern for a bracket as required for the ends of winders Upon the top edge of the bracket used for the flyers describe an equilateral triangle. Divide the contour of the bracket into a number of parts, and draw lines from divisions perpendicular to the top or base of the triangle.

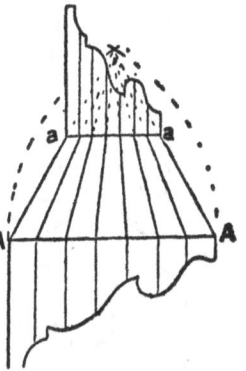

Fig. 92.

From these intersections draw lines to the apex of the triangle. Next mark upon the sides of the triangle, from the apex, the length of the bracket required. Join these points by a line, *a a*, which is parallel with the base, and upon the points where the line cuts the lines drawn to the apex, erect perpendiculars; make them equal in length to the corresponding lines drawn on the original bracket.

The eight illustrations shown in Fig. 93 give brackets and sections of handrails of various kinds, and is

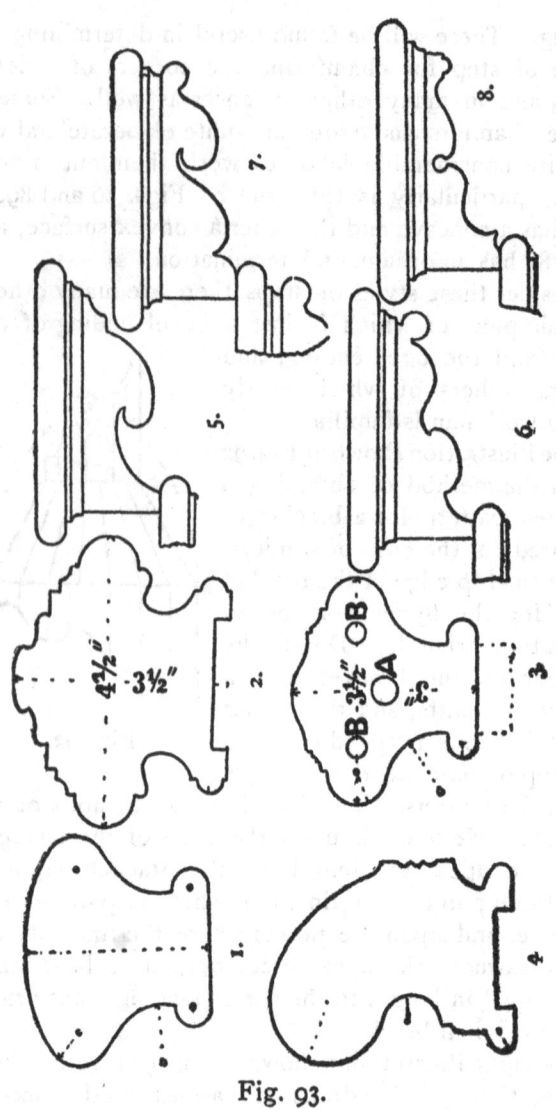

Fig. 93.

FOURTH METHOD

offered as a supplement to the page of handrail sections shown in part three of this work. In the examples given the centers of the curves forming the handrails are given, and the sizes of the rails are marked on the sections in two instances. The numbers 5 to 8, inclusive, show patterns for brackets which may be made to suit almost any style of stairs. Other patterns will be found illustrated in previous pages of this work in connection with examples of platform stairs.

TABLES

The following tables which are taken from the *Builder and Woodworker*, but which I believe were first prepared by *The California Architect*, will be found very useful to those "figuring" on the run and rise of stairs. The spacing of the lines of figures into groups aids the eye in following the direction to the final point.

Directions:—In the column beginning with the rise of step desired, find the height of story from top of floor to top of floor, then follow this line to the column under risers, which gives the number of risers. In the column under "treads" find the number of risers, *less one*, and on this line under the column of width of tread will be the length of run.

Risers.	Inches.	Inches.	Inches.	Inches.	Inches.	Inches.	Inches.	Inches.	Inches.	Inches.	Inches.
No. 1......	0.6	0. 6¼	0. 6½	0. 6¾	0. 7	0. 7⅛	0. 7¼	0. 7⅜	0. 7½	0. 7⅝	0. 7¾
2......	1.0	1. 0½	1. 1	1. 1½	1. 2	1. 2¼	1. 2½	1. 2¾	1. 3	1. 3¼	1. 3½
3......	1.6	1. 6¾	1. 7½	1. 8¼	1. 9	1. 9⅜	1. 9¾	1.10⅛	1.10½	1.10⅞	1.11¼
4......	2.0	2. 1	2. 2	2. 3	2. 4	2. 4½	2. 5	2. 5½	2. 6	2. 6½	2. 7
5......	2.6	2. 7¼	2. 8½	2. 9¾	2.11	2.11⅝	3. 0¼	3. 0⅞	3. 1½	3. 2⅛	3. 2¾
6......	3.0	3. 1½	3. 3	3. 4½	3. 6	3. 6¾	3. 7½	3. 8¼	3. 9	3. 9¾	3.10½
7......	3.6	3. 7¾	3. 9½	3.11¼	4. 1	4. 1⅞	4. 2¾	4. 3⅝	4. 4½	4. 5⅜	4. 6¼
8......	4.0	4. 2	4. 4	4. 6	4. 8	4. 9	4.10	4.11	5. 0	5. 1	5. 2
9......	4.6	4. 8¼	4.10½	5. 0¾	5. 3	5. 4⅛	5. 5¼	5. 6⅜	5. 7½	5. 8⅝	5. 9¾
10......	5.0	5. 2½	5. 5	5. 7½	5.10	5.11¼	6. 0½	6. 1¾	6. 3	6. 4¼	6. 5½
11......	5.6	5. 8¾	5.11½	6. 2¼	6. 5	6. 6⅜	6. 7¾	6. 9⅛	7.10½	6.11⅞	7. 1¼
12......	6.0	6. 3	6. 6	6. 9	7. 0	7. 1½	7. 3	7. 4½	7. 6	7. 7½	7. 9
13......	6.6	6. 9¼	7. 0½	7. 3¾	7. 7	7. 8⅝	7.10¼	8.11⅞	8. 1½	8. 3⅛	8. 4¾
14......	7.0	7. 3½	7. 7	7.10½	8. 2	8. 3¾	8. 5½	8. 7¼	8. 9	8.10¾	9. 0½
15......	7.6	7. 9¾	8. 1½	8. 5¼	8. 9	8.10⅞	9. 0¾	9. 2⅝	9. 4½	9. 6⅜	9. 8¼
16......	8.0	8. 4	8. 8	9. 0	9. 4	9. 6	9. 8	9.10	10. 0	10. 2	10. 4
17......	8.6	8.10¼	9. 2½	9. 6¾	9.11	10. 1⅛	10. 3¼	10. 5⅜	10. 7½	10. 9⅝	10.11¾
18......	9.0	9. 4½	9. 9	10. 1½	10. 6	10. 8¼	10.10½	11. 0¾	11. 3	11. 5¼	11. 7½
19......	9.5	9.10¾	10. 3½	10. 8¼	11. 1	11. 3⅜	11. 5¾	11. 8⅛	11.10½	12. 0⅞	12. 3¼
20......	10.0	10. 5	10.10	11. 3	11. 8	11.10½	12. 1	12. 3½	12. 6	12. 8½	12.11
21......	10.6	10.11¼	11. 4½	11. 9¾	12. 3	12. 5⅝	12. 8¼	12.10⅞	13. 1½	13. 4⅛	13. 6¾
22......	11.0	11. 5½	11.11	12. 4½	12.10	13. 0¾	13. 3½	13. 6¼	13. 9	13.11¾	14. 2½
23......	11.6	11.11¾	12. 5½	12.11¼	13. 5	13. 7⅞	13.10¾	14. 1⅝	14. 4½	14. 7⅜	14.10¼
24......	12.0	12. 6	13. 0	13. 6	14. 0	14. 3	14. 6	14. 9	15. 0	15. 3	15. 6
25......	12.6	13. 0¼	13. 6½	14. 0¾	14. 7	14.10⅛	15. 1¼	15. 4⅜	15. 7½	15.10⅝	16. 1¾
26......	13.0	13. 6½	14. 1	14. 7½	15. 2	15. 5¼	15. 8½	15.11¾	16. 3	16. 6¾	16. 9½
27......	12.6	14. 0¾	14. 7½	15. 2¼	15. 9	16. 0⅜	16. 3¾	16. 7⅛	16.10½	17. 1⅞	17. 5¼
28.. .	14.0	14. 7	15. 2	15. 9	16. 4	16. 7½	16.11	17. 2½	17. 6	17. 9½	18. 1

FOURTH METHOD

Treads.	Inches	Inches.	Inches.	Inches.	Inches.	Inches.	Inches.	Inches.	Inches.	Inches.	Inches.
1	0.14	0.13	0.11	0.10½	0.10	0. 9½	0.9	0. 8½	0. 8¼	0.8	0. 7⅞
2	2. 4	2. 2	1.10	1. 9	1. 8	1. 7	1.6	1. 5	1. 4½	1.4	1. 3¾
3	3. 6	3. 3	2. 9	2. 7½	2. 6	2. 4½	2.3	2. 1½	2. 0¾	2.0	1.11⅜
4	4. 8	4. 4	3. 8	3. 6	3. 4	3. 2	3.0	2.10	2. 9	2.8	2. 7½
5	5.10	5. 5	4. 7	4. 4½	4. 2	3.11½	3.9	3. 6½	3. 5¼	3.4	3. 3⅜
6	7. 0	6. 6	5. 6	5. 3	5. 0	4. 9	4.6	4. 3	4. 1½	4.0	3.11¼
7	8. 2	7. 7	6. 5	6. 1½	5.10	5. 6½	5.3	4.11½	4. 9¾	4.8	4. 7⅛
8	9. 4	8. 8	7. 4	7. 0	6. 8	6. 4	6.0	5. 8	5. 6	5.4	5. 3
9	10. 6	9. 9	8. 3	7.10½	7. 6	7. 1½	6.9	6. 4½	6.10½	6.0	5.10⅞
10	11. 8	10.10	9. 2	8. 9	8. 4	7.11	7.6	7. 1	6.10¾	6.8	6. 6¾
11	12.10	11.11	10. 1	9. 7½	9. 2	8. 8½	8.3	7. 9½	7. 6¾	7.4	7. 2⅝
12	14. 0	13. 0	11. 0	10. 6	10. 0	9. 6	9.0	8. 6	8. 3	8.0	7.10½
13	15. 2	14. 1	11.11	11. 4½	10.10	10. 3½	9.9	9. 2½	8.11¼	8.8	8. 6⅜
14	16. 4	15. 2	12.10	12. 3	11. 8	11. 1	10.6	9.11	9. 7½	9.4	9. 2¼
15	17. 6	16. 3	13. 9	13. 1½	12. 6	11.10½	11.3	10. 7½	10. 3¾	10.0	9.10⅛
16	18. 8	17. 4	14. 8	14. 0	13. 4	12. 8	12.0	11. 4	11. 0	10.8	10. 6
17	19.10	18. 5	15. 7	14.10½	14. 2	13. 5½	12.9	12. 0½	11. 8¼	11.4	11. 1⅞
18	21. 0	19. 6	16. 6	15. 9	15. 0	14. 3	13.6	12. 9	12. 4½	12.0	11. 9¾
19	22. 2	20. 7	17. 5	16. 7½	15.10	15. 0½	14.3	13. 5½	13. 0¾	12.8	12. 5⅝
20	23. 4	21. 8	18. 4	17. 6	16. 8	15.10	15.0	14. 2	13. 9	13.4	13. 1½
21	24. 6	22. 9	19. 3	18. 4½	17. 6	16. 7½	15.9	14.10½	14. 5¼	14.0	13. 9⅜
22	25. 8	23.10	20. 2	19. 3	18. 4	17. 5	16.6	15. 7	15. 1½	14.8	14. 5¼
23	26.10	24.11	21. 1	20. 1½	19. 2	18. 2½	17.3	16. 3½	15. 9¾	15.4	15. 1⅛
24	28. 0	26. 0	22. 0	21. 0	20. 0	19. 0	18.0	17. 0	16. 6	16.0	15. 9
25	29. 2	27. 1	22.11	21.10½	20.10	19. 9½	18.9	17. 8½	17. 2¼	16.8	16. 4⅞
26	30. 4	28. 2	23.10	22. 9	21. 8	20. 7	19.6	18. 5	17.10½	17.4	17. 0¾
27	31. 6	29. 3	24. 9	23. 7½	22. 6	21. 4½	20.3	19. 1½	18. 6¾	18.0	17. 8⅝
28	32. 8	30. 4	25. 8	24. 6	23. 4	22. 2	21.0	19.10	19. 3	18.8	18. 4½

A GLOSSARY OF TERMS USED IN STAIR-BUILDING AND HAND-RAILING

A Flight.—A continued series of steps without a break.

Axis.—In *architecture* an imaginary line through the center of a column, etc., or its geometrical representation; where different members are placed over each other so that the same vertical line on the elevation divides them equally, they are said to be on the same axis, although they may be on different planes; thus triglyphs and modillions are so arranged that one coincides with the axis or line of axis of each column; in like manner the windows or other openings in the several stories of a façade must all be in the same respective axis whether they are all of the same breadth or not. In *geometry*, the straight line in a plane figure about which it revolves to produce or generate a solid. In *mechanics*, the axis of a balance is the line upon which it moves or turns. In *turning*, an imaginary line passing longitudinally through the middle of the body to be turned, from one point to the other of the two cones, by which the work is suspended or between the back center and the center of the collar of the puppet which supports the end of the mandril at the chuck.

GLOSSARY

Baluster.—A small column or post turned of different forms and sizes, forming an ornamental enclosure and supporting the handrail; generally two to a step.

Balustrade.—A series or row of balusters joined by a rail, serving for a rest to the arms, or as a fence or enclosure to balconies, altars, staircases, etc. Balustrades when intended for use or against windows or flights of steps, terraces and the like, should not be more than 3 feet 6 inches, nor less than 3 feet in height. When used for ornament, as on the summit of a building, their height may be from two-thirds to four-fifths of the entablature whereon they are employed, and this proportion is to be taken exclusive of their zoccolo or plinth, so that from the proper point of sight the whole balustrade may be exposed to view. There are various species of balusters; if single-bellied the best way is to divide the total height of the space allotted for the balustrade into thirteen equal parts, the height of the baluster to be eight, of the base three, and of the cornice two of those parts; or divide the total height into fourteen parts, making the baluster eight, the base four, and the cornice two. If double-bellied the height should be divided into fourteen parts, two of which are to be given to the cornice, three to the base and the remainder to the baluster.

The distance between two balusters should not be more than half the diameter of the baluster in

GLOSSARY

its thickest part, nor less than one-third of it; but on inclined planes the intervals should not be quite so wide.

Butt Joint.—An end joint made at right angles to the central tangent of a wreath piece; and also an end joint made at right angles to any straight length of handrail.

Carriage.—The timber work which supports the steps of a wooden stair.

Close String.—In dog-leg stairs a staircase without an open newel.

Cockel Stairs.—A winding staircase.

Circular Stairs are stairs with steps planned in a circle toward the center of which they all converge and are all winders.

Curve-out.—A concave curve of the face of a front-string at its starting.

Curtail Step.—The first step by which a stair is ascended, finishing at the end in a form of a scroll following the plan of the handrail.—*Nicholson.*

Cylinder.—A *cylinder* is a solid described by geometricians as generated by the rotation of a rectangle about one of its sides supposed to be at rest; this quiescent side is called the *axis* of the *cylinder*, therefore the base and top of the cylinder are equal or similar circles.

A *prism* is a solid, whose base and top are similar right line figures, with sides formed in planes, and rising perpendicularly from the base to the top.

The *cylinder*, so called by *joiners*, is a solid figure compounded of the two last mentioned figures; its base is composed of a *semicircle* joined to a *right-angled* parallelogram. This last compound figure is intended whenever the word *cylinder* occurs in the preceding work unless the word geometrical be prefixed.

Dog-legged Stairs.—Such as are solid between the upper flights, or those that have no well-hole; and the rail and balusters of both the progressive and retrogressive flight fall in the same vertical plane. The steps are fixed to strings, newels and carriages; and the ends of the steps of the inferior kind terminate *only* on the side of the string.—*Nicholson*.

Elliptic Stairs.—Stairs that are elliptic on the plan, the treads all converging, but not to one center like those of a circular stair.

Face Mould.—A section produced on any inclined plane vertically over a curved plan of handrail.

Flight of Stairs.—In a staircase the series of steps from one landing place to another. Thus the same staircase between one floor and another may consist of more than one flight of steps, the flight being reckoned from one landing to another.

Front String.—The string on that side of the stair over which the handrail is placed.

Fillet.—A band 1¼ inches wide by ¼ inch thick nailed to the face of a front string below the cove and extending the width of a tread.

GLOSSARY

Flyers.—Steps in a flight that are parallel to each other.

Geometrical Stair.—A flight of stairs supported only by the wall at one end of the steps.

Half-space, or resting place.—The interval between two flights of steps in a staircase.

Hall.—The first large apartment on entering a house; the public room of a corporative body; a manor-house.

Handrail.—A variously formed and sized rail running parallel to the inclination of the stairs for holding the balusters.

Hollow Newel.—An opening in the middle of the staircase. The term is used in contradistinction to *solid newel*, into which the ends of the steps are built. In the hollow newel, or well-hole, the steps are only supported at one end by the surrounding wall of the staircase, the ends next the hollow being unsupported.—*Nicholson*.

Helix.—The small twist under the head of a Corinthian column.

Housing.—The space excavated out of a body, for the insertion of some part of the extremity of another in order to fasten the two together; thus the string-board of a stair is most frequently excavated, or notched out for the reception of steps. The term is also applied to a niche for containing a statue.—*Nicholson*.

Joint.—The surface of separation between two bodies brought into contact and held firmly together, either by some cementing medium or by the weight of one body lying upon another. A joint is not merely the contact of two surfaces, though the nearer they approach the more perfect the joint. In masonry the distance of the planes intended to form a joint is comparatively considerable because of the coarseness of the particles which enter into the composition of the cement.

Kerf.—A slit or cut in a piece of timber or in a stone, usually applied to that made by a saw or axe.

Keys.—In naked flooring, pieces of timber fixed in between the joists by mortise and tenon; when these are fastened with their ends projecting against sides they are termed strutting pieces.

Keys.—Pieces inserted in boards to prevent warping.

Knee.—A convex bend in the back of a handrail.

Knee.—A part of the back of a handrailing of a convex form, the reverse of a *ramp*, which is a back of a handrail and is concave; also any piece of timber bent to an angular joint.

Landing.—Horizontal resting-place in a flight.

Newel.—The central column around which the steps of a circular staircase wind; the principal post at the angles and foot of a staircase.

Newel.—In architecture the upright post or central column around which the steps of a circular staircase are made to wind, being that part of the staircase by which they are sustained.

The newel is properly a cylinder of stone or

wood, which bears on the ground and is formed by the ends of the steps of the winding stairs.

There are also newels of wood, which are pieces of wood placed perpendicularly, receiving the tenons of the steps of wooden stairs into their mortises, and wherein are fitted the shafts and rests of the staircase and the flight of each story. In some of the Tudor and Elizabethan residences some very fine examples may be seen of the newel richly ornamented and adding much to the beauty of the staircase.—*Nicholson.*

Nosing.—The outer or front edge of the step.

Pitching Piece.—A horizontal timber with one of its ends wedged into the wall at the top of a flight of stairs to support the upper end of the rough strings.

Pitch.—Angle of inclination of the stairs.

Pitch-board.—A piece of thin board in the form of a right-angled triangle, one of the sides of the right angle equal to a rise.

Ramp.—A concave or convex curve or casement of an angle, as sometimes required at the end of a wreath or an adjoining straight rail.

Rise.—The vertical rise between the treads.

Riser.—The board forming the vertical portion of the front of a step.

Run.—Of a flight of stairs, the horizontal distance from the first to the last riser.

Scroll.—A carved curvilinear ornament, somewhat

resembling in profile the turnings of a ram's horn. —*Hatfield*.

Splay.—A slanting or beveling in the sides of an opening to a wall for a window or door, so that the outside profile of the window is larger than that of the inside; it is done for the purpose of facilitating the admission of light. It is a term applied to whatever has one side making an oblique angle with the other; thus the heading joists of a boarded floor are frequently splayed in their thickness. The word *fluing* is sometimes applied to an aperture in the same sense as *splayed*.

Spring Bevel of a Rail.—The angle made by the top of the plank with a vertical plane touching the ends of the rail piece which terminates the concave side.

Squaring a Handrail.—The method of cutting a plank to the form of a rail for a staircase so that all the vertical sections may be right angles.

Spiral.—In *geometry*, a curve line of the circular kind, which in its progress always recedes more and more from its center. In *architecture*, a curve that ascends winding about a cone or spire so that all its points continually approach its axis.

Spandril.—The angle formed by a stairway.

Stairs (from the Saxon *stæger*).—In a building, the steps whereby to ascend and descend from one story to another.

GLOSSARY

The breadth of the steps of stairs in general use in common dwelling houses is from 9 to 12 inches, or about 10 inches medium. In the best staircases of fine houses or public edifices the breadth ought never to be less than 12 inches nor more than 18 inches. It is a general maxim that the greater breadth of a step requires less height than one of less breadth; thus, a step of 12 inches in breadth will require a rise of $7\frac{1}{2}$ inches, which may be taken as a standard by which to regulate those of other dimensions; so that multiplying 12 inches by $5\frac{1}{2}$ we shall have 66; then supposing a step to be 10 inches in breadth the height should be $66 - 10 = 6\frac{5}{8}$ inches, which is nearly, if not exactly, what common practice would allow. The proportion of steps being thus regulated the next consideration is the number requisite between two floors or stories which will be ascertained by supposing the breadth of the steps given, say 10 inches each, as depending on the space allowed for the staircase, and this, according to the rule laid down, will require a rise of nearly 7 inches; suppose then the distance from floor to floor to be 13 feet 4 inches, or 160 inches, $160 \div 7 = 22\frac{6}{7}$, which would be the number required; but as all the steps must be of equal heights we should rather take 23 risers, provided the staircase room would allow it, and so make the height of each somewhat less than 7 inches.

The most certain method of erecting a staircase is to provide a rod of sufficient length to reach from one floor to another, divided into as many equal parts as the intended number of risers, and try every step as it is set to its exact height. The breadth of the staircase may be from 6 to 20 feet according to the use or application of the building or the form or proportions of the plan.

If the steps be less than 3 feet in length the staircase becomes inconvenient for the passing of furniture, as is frequently the case in small houses.

Though it is desirable to have such rules as are here laid down for regulating the proportion of the heights, breadths and lengths of steps, architects and workmen cannot be so strictly tied to them but that they may vary them as circumstances may demand.—*Nicholson.*

Stairs are constructions composed of horizontal planes elevated above each other, forming steps, affording the means of communication between the different stories of a building.

In the distribution of a house of several stories the stairs occupy an important place. In new constructions their form may be regular, but in the reparation or remodeling of old buildings the first consideration is generally to make the distribution suitable for the living and sleeping rooms, and then to convert to the use of the stairs the spaces which may remain, giving to them such forms in plan as will render them agreeable to the sight and commodious in the use.

When houses began to be built in stories the stairs were placed from story to story in straight flights like ladders. They were erected on the exterior of the building, and to shelter them when so placed great projection was given to the roofs. To save the extent of space required by straight flights the stairs were made to turn upon themselves in a spiral form, and were inclosed in turrets. A newel, either square or round, reaching from the ground to the roof, served to support the inner ends of the steps, and the outer ends were let into the walls or supported on notched boards attached to the walls.

At a later period the stairs came to be inclosed within the building itself, and for a long time preserved the spiral form which the former situation had necessitated.

Definitions.—The apartment in which the stair is placed is called the *staircase*.

The horizontal part of a step is called the *tread*, the vertical part the *riser*, the breadth or distance from riser to riser the *going*, the distance from the first to the last riser in a flight the *going of the flight*.

When the risers are parallel with each other the stairs are, of course, *straight*.

When the steps are narrower at one end than the other they are termed *winders*.

When the bottom step has a circular end it is called a *round-ended step;* when the end is formed into a spiral it is called a *curtail step.*

The wide step introduced as a resting place in the ascent is a *landing*, and the top of a stair is also so called.

When the landing occupies the whole width of the staircase it is called a *half-space.*

When the landing at a resting place is square it is designated a *quarter-space.*

So much of a stair as is included between two landings is called a *flight*, especially if the risers are parallel with each other; the steps in this case are *fliers.*

The outward edge of a step is named the *nosing;* if it projects beyond the riser so as to receive a hollow moulding glued under it it is a *moulded nosing.*

A straight edge laid on the nosings represents the angle of the stairs, and is denominated the *line of nosings.*

The raking pieces which support the ends of the steps are called *strings*. The inner one placed against the wall is the *wall string;* the other the *outer string.* If the outer string be cut to miter with the end of the riser it is a *cut and mitered string;* but when the strings are grooved to receive the ends of the treads and risers they are said to be *housed*, and the grooves are termed *housings.*

Stairs in which the outer string of the upper flight stands perpendicularly over that of the lower flight are called *dog-legged stairs*, otherwise *newel stairs*, from the fact of a piece of stuff called a *newel*, being used as the axis of the spiral of the stair; the newel is generally ornamented by turning, or in some other way. The outer strings in such stairs are tenoned into the newel, as also are the first and last risers of the flight.—*Newland*.

Staircase.—A term applied to the whole set of stairs, with the walls supporting the steps, leading from one story to another. The same staircase frequently conducts to the top of the building, and thus consists of as many stories as the building itself.

When the height of the story is considerable, resting places become necessary, which go under the name of *quarter-spaces* and *half-spaces*, according as the passenger has to pass a right angle, or two right angles; that is, as he has to describe a quadrant or a semicircle. In very high stories that admit of sufficient head-room, and where the space allowed for the staircase is confined, the staircase may have two revolutions in the height of one story, which will lessen the height of the steps; but in grand staircases only one revolution can be admitted, the length and breadth of the space on the plan being always proportioned to the height of the building, so as to admit of fixed proportions.

In contriving a grand edifice particular attention must be paid to the situation of the space occupied by the stairs, so as to give them the most easy command of the rooms.

With regard to the lighting of a grand staircase, a skylight, or rather lantern, is the most appropriate; for the light thus admitted is powerful, and the design admits of greater elegance; indeed, where the staircase does not adjoin the exterior walls this is the only method by which light can be admitted.

In small buildings the position of the staircase is indicated by the general distribution of the plan, but in large edifices this is not so obvious, but must at least be determined by considering naturally its connection with other apartments.— *Nicholson.*

Straight Flight of Stairs is one in which the steps are parallel and at right angles to the strings.

Steps (from the Saxon *stæp*).—The degrees of a staircase, by which we rise, consisting of two parts, one horizontal called *treads*, the other vertical called *risers*. When steps are placed around the circumference of a circle, or an ellipse, or any segments of them, they are called *winders*, but when the sides are straight they are called *fliers*. The first, or lower step, with a scroll wrought upon its end, according to the plan of the handrail, is called *the curtail step*.

GLOSSARY

Stretch-out.—A term applied to a surface that will just cover a body so extended that all its parts are in a plane, or may be made to coincide with a plane.

Scroll or Curtail Step.—The bottom step with the front end shaped to receive the balusters around the scroll of the handrail.

String or String-piece.—That part of a flight of stairs which forms its ceiling or soffit.

String-board.—In wooden stairs the board next the well-hole which receives the ends of the steps; its face follows the direction of the well-hole, whatever the form; when curved it is frequently formed in thicknesses glued together, though sometimes it is got out of the solid like a handrail.

String-board.—In wooden stairs, a board placed next to the well-hole, and terminating the ends of the steps. The face of string-boards follows the direction of the well-hole, whether it be prismatic or an inverted cone. String-boards are sometimes glued in several thicknesses with the fibers of the wood running in the direction of the steps; sometimes they are wrought out of solid, like a handrail, the grain of the wood being in the same direction; and they are also glued up like columns, viz., having the fibers vertical. Brackets are most frequently placed upon the string-boards and mitered into the risers.—*Nicholson.*

Tangent.—In *geometry*, a right line perpendicularly raised on the extremity of a radius, which touches a circle so that it would never cut it, although indefinitely produced, or in other words, it would never come within its circumference.

Step.—The horizontal board on which we tread.

Soffit.—The under side of an arch or moulding.

Tread.—The horizontal distance between the risers—one of the equal divisions into which the flight is divided; the top of the step.

Wall String.—The board placed against the wall to receive the ends of the step.

Well.—The place occupied by the flight of stairs. The space left beyond the ends of the steps is called the well-hole.

Well Staircase.—A winding staircase of ascent or descent, to different parts of a building, so called from the walls inclosing it resembling a well; called frequently a geometrical staircase.

Winders.—Stairs, steps not parallel to each other.

The winders are supported by rough pieces called *bearers*, wedged into the wall and secured to the strings.

When the front string is ornamented with brackets it is called a *bracketed stair*.

Treads of triangular form used to turn an angle or go round a curve.

Wreath.—The whole of a heliacally curved handrail.

Wreath Piece.—A portion of a wreath less than the whole.

CONTENTS
FIRST METHOD

	Page
Preface	5
Advice to Young Workmen	9
Straight Flight of Stairs	11
Landing Stairs	11
Acute Landing and Cylinder	12
Obtuse Landing and Cylinder	12
Half-space and Two-step Landing	12
Quarter-space and Four Winders	13
Quarter-space and Six Winders	13
Half-space and Dancing Winders	14
Half-space, Cylinder and Dancing Winders	14
Circular Stairs	15
Diagram of Tangents	16
Curve of Face Mould	17
Explanatory Diagram	18
Rule for Turn-out	20
Steps and Risers	20
Line of Rail	21
Face Moulds	23
Facing Mould Lines	24
Acute Angle Stairs	26
Showing Mould and Pitch	27
Blocking Out	28
Constructing Cylinder	29
Stretch-out	30

CONTENTS

	Page
Getting Bevels for Butts	31
Face Moulds and Stretch-outs	32
Ramps over Fliers	33
Moulds for Quarter-space Stairs	34
Stretch-out over Cylinder	36
Stretch-out over Winders	38
Laying Out Rail over Circular Well-hole	40
Face Moulds, Ramps and Stretch-outs	41
Final Remarks	42

SECOND METHOD

A Remark or Two	43
Given Treads and Risers	44
Use of Pitch-board	46
Stair Strings and Winders	47
Line Theory of Handrailing	49
Around a Cylinder	50
How to Obtain a Wreath	52
Squaring a Wreath	54
Twists and Cylinders	56
Cutting Wreath Square to Plank	57
Beveling Joints	58

THIRD METHOD

Line of Nosings	59
Rail over Level Landing	59
Face Moulds, Tangents and Joints	60
Center Lines for Rails	60
Major and Minor Axes	61
Blocking Out for Wreath	62

CONTENTS

	Page
Rail in Position	63
Rail at Landing	64
Line of Quadrant	65
Lay-out of Pitches	66
Pitch in Cylinder, How Found	70
Bevels on Cylinder	71
Face Mould	72
Blocking Out	74
Plan of Quarter-space Rail	75
Wreath Worked Out	77
Tangents Unfolded	79
Ramp and Pitch	80
Stretch-out for Winders	81
Ramp and Templets	83
Method of Getting Face Moulds	84
Plan of Wreath, Risers and Tangents	85
Stretch-outs over Landing	86
Isometrical Sketch of Wreath	87
Stretch-out over Obtuse Landing	89
Wreath over Cylinder	90
Bevels for Butt Cuts	91
Falling Lines and Well	92
Use of Pitch-board	93
Wreath for Small Cylinder	94
Three Points in a Cylinder	96
Sliding the Face Moulds	96
Mould and Plank	97
Nine Sections of Handrails	99
General Glossary and Definitions	100

CONTENTS

FOURTH METHOD

	Page
A Philadelphia Stairway	145
A Sensible Stairway	146
A Continued Stair	149
A Quaint Stair	150
A Built-Up Newel	155
Built-Up Newels	160
Balusters with Brass Brackets	165
Cutting Miter Cap	114
Cutting Strings	115
Carriages for Strings	121
Carved Stairway	138
Carved Newel Top	140
Colonial Stairway	143
Carved Newel from Belgium	157
Columner Newel	161
Chamfering	167
Dovetailed Treads	119
Double Stairs	131
Details of Carved Stairway	139
Details of Rail and Finish	152
Difference in Balusters	164
Diminishing Brackets	169
Elevation of Stairs	111
Elevation of Housed String	117
Ends of Steps	123
Elevation of Bull-Nose Steps	125
Elevation of Stairway	135
Elevation of Grand Entrance	136
Framing Platform Stairs	108
Framing a Bull-Nose Step	112
Four Plans of Stairs	132
Fancy Stair and Newel	156

CONTENTS

	Page
Grand Stairs with Landings	131
Grand Hall Stairs	131
Head Room	128
Heavy Carved Newels	141
Introduction to Method IV	99
Interior View of Stairway	153
Newels and Platform Stairs	99
Newel Cap	113
Newel Miter Cap	114
Nosings	121
Nosing Solid	122
Newel and Baluster	154
Open Strings	109
Open Newel Stairs	110
Open String and Nosing	125
Outside Hall Plan	135
Odd Newels from France	161
Ornamental Stop Chamfering	168
Proportioning Treads and Risers	100
Platforms	107
Plan of Stairs	110
Plan of Tread and Riser	121
Plan of Showing Carriage Strings	126
Perspective of Open Stairs	126
Perspective of Carved Newel	142
Plain Newels	158
Pitman's Carved Newel	159
Patterns of Handrails	170
Rail and Baluster Fastenings	123
Raking and Straight Balusters	147
Strength of Stairs	107
Sketches of Cut Strings	118
Section of Bull-Nose Step	124
Section of Rail with Balusters	148

CONTENTS

	Page
Serpentine Newel and Balusters	151
Sections and Elevations of Stair	152
Some French Newels	162
Spanish Newels	163
Spiral Newel	165
Stop Chamfering	167
Styles of Brackets	170
Sections of Handrails	170
Two Landing Stairs	129
Three Landing Stairs	130
Three Plans of Stairs	133
Three Hall Plans of Stairs	134
Two Elevations of Stairs	166
Tables of Treads and Risers	171

HOUSE PLAN SUPPLEMENT

PERSPECTIVE VIEWS AND FLOOR PLANS

OF

Twenty-five Low and Medium Priced Houses

Full and Complete Working Plans and Specifications of any of these houses will be mailed at the low prices named, on the same day the order is received

OTHER PLANS

We illustrate in "Practical Uses of the Steel Square," Vol I, "Practical Uses of the Steel Square," Vol II, and "Modern Carpentry," 75 other plans, 25 in each book, none of which are duplicates of those we illustrate herein

For further information, address

THE PUBLISHERS

Send All Orders for Plans to

The RADFORD ARCHITECTURAL COMPANY

CHICAGO, ILLINOIS · 192 West 22d Street
RIVERSIDE, ILLINOIS   Green Block

25 HOUSE DESIGNS 25

Without extra cost to our readers, we have added to this volume the perspective view and floor plans of twenty-five low and medium priced houses, such as 90 per cent of the home builders to-day wish to build. In the drawing of these plans special effort has been made to provide for the most economical construction, thereby giving the home builder and the contractor the benefit of the saving of many dollars, for in no case have we put any useless expense upon the building, simply to carry out some pet idea. Every plan illustrated will show by the complete working plans and specifications that we give you designs that will work out to the best advantage and will give you the most for your money, besides, every bit of space has been utilized to the best advantage.

This supplement, as well as all other books published by this company, has for its foundation the best equipped architectural establishment ever maintained for the purpose of furnishing the public with complete working plans and specifications at the remarkably low price of only $5.00 per set. Every plan is designed by a licensed architect, who stands at the head of his profession in this particular class of work. The Radford Houses are now being erected in every country of the world where frame houses are built, which bespeaks for our plans more than anything we can say.

What We Give You

The first question you will ask is, "What do we get in these complete working plans and specifications? Of what do they consist? Are they the cheap, printed plans on tissue paper without details or specifications?"

We do not blame you for wishing to know what you will get for your money. The plans we send out are the regular blue-printed plans drawn one-quarter inch scale to the foot, showing all the elevations, floor plans and necessary interior details. We use the very best quality heavy Gallia Blue Print Paper, number 1000-X, using great care in the blue-printing to have every line and figure perfect and distinct.

What We Furnish in Blue Prints

Foundation and Cellar Plan

This sheet shows the shape and size of all walls, piers, footings, posts, etc., and of what materials they are constructed; shows the location of all windows, doors, chimneys, ash pits, partitions and the like. The different wall sections are given, showing their construction and measurements from all the different points.

Floor Plans

These plans show the shape and size of all rooms, halls, and closets; the location and size of all doors and windows, the position of all plumbing fixtures, gas lights, registers, pantry work, etc., and all the measurements that are necessary are given.

Elevations

A front, right, left and rear elevation are furnished with all the plans. These drawings are complete and accurate in every respect. They show the shape, size and location of all doors and windows, porches, cornices, towers, bays and the like, and, in fact, give you an exact scale picture of the house as it should be at completion. Full wall sections are given, showing the construction from foundation to roof, the height of stories between the joists, height of plates, pitch of roof, etc.

Roof Plan

This plan is furnished where the roof construction is at all intricate. It shows the location of all hips, valleys, ridges, decks, etc.

All the above drawings are made to scale one-quarter inch to the foot

Details

All necessary details of the interior work, such as door and window casings and trim, base, stools, picture moulding, doors, newel posts, balusters, rail, etc , accompany each set of plans. Part is shown in full size, while some of the larger work, such as stair construction, is drawn to a scale of one and one-half inch to the foot.

These blue prints are substantially and artistically bound in cloth and heavy waterproof paper, making a handsome and durable covering and protection for the plans

Specifications

The specifications are typewritten on Lakeside Bond Linen Paper and are bound in the same artistic manner as the plans, the same cloth and waterproof paper being used. They consist of from about sixteen to twenty pages of closely typewritten matter, giving full instructions for carrying out the work. All directions necessary are given in the clearest and most explicit manner, so that there can be no possibility of a misunderstanding.

Basis of Contract

These working plans and specifications can be made the basis of contract between the home builder and the contractor. They will prevent mistakes which cost money, and they will prevent disputes which are unforeseen and never settled satisfactorily to both parties When no plans are used, the contractor is often obliged to do some work which he did not figure on, and the home builder often does not get as much for his money as he expected, simply because there was no basis on which to work and upon which to base the contract

No misunderstandings can arise when a set of our plans and specifications are before the contractor and the home builder, showing the interior and exterior construction of the house as agreed upon in the contract. Many advantages may be claimed for the complete working plans and specifications. They are time savers, and therefore money savers Workingmen will not have to wait for instructions when a set of plans is left on the job. They will prevent mistakes in cutting lumber, in placing door and window frames, and in many other places when the contractor is not on the work and the men have received only partial or indefinite instructions. They also give instructions for the working of all material to the best advantage.

Free Plans for Insurance Adjustment

You take every precaution to have your house covered by insurance, but do you make any provision for the adjustment of the loss should you have a fire? There is not one man in ten thousand who will provide for this embarrassing situation. You can call to mind instances in your own locality where settlements have been delayed because the insurance companies wanted some proof which could not be furnished.

They demand proof of loss before paying insurance money, and they are entitled to it. We have provided for this and have inaugurated the following plan, which cannot but meet with favor by whoever builds a house from our plans:

Immediately upon receipt of information from you that your house has been destroyed by fire, either totally or partially, we will forward you, free of cost, a duplicate set of plans and specifications, and in addition we will furnish an affidavit giving the number of the design and date when furnished, to be used for the adjustment of the insurance.

Without one cent of cost to you and without one particle of trouble, we keep a record of the number of the house design and the date it was furnished, so that, in time of loss, all it will be necessary for you to do is to drop us a line and we will furnish the only reliable method of getting a speedy and satisfactory adjustment. This may be the plans for saving you hundreds of dollars besides much time and worry.

Our Liberal Prices

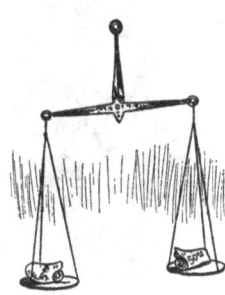

Many have marveled at our ability to furnish such excellent and complete working plans and specifications at such low prices. We do not wonder at this, because we charge but five dollars for a more complete set of working plans and specifications than you would receive if ordered in the regular manner, and when drawn especially for you, at a cost of from fifty to seventy-five dollars. On account of our large business and unusual equipment, and owing to the fact that we divide the expense of these plans among so many, it is possible for us to sell them at these low prices. The margin of profit is very close, but it enables us to sell thousands of sets of plans, which save many times their cost to both the owner and the contractor in erecting even the smallest dwelling.

Our Reliability

Our reliability is beyond question. We have been in the business for many years, having grown from a small institution to our present large capacity, publishing many books and furnishing plans and specifications for many thousands of houses in all parts of the United States, Canada, Europe, Australia, and South Africa. We presume this book may fall into the hands of some one who does not know us, therefore, if you have never heard of us and are not familiar with our reliability and business methods, inquire of your lumber dealer or banker. This article is unnecessary to those who have had previous dealings with us If you are afraid to send the money direct to us, send it with your order to The Federal Trust and Savings Bank, of Chicago, Ill. (capital and surplus, $2,500,000), or to the Riverside State Bank of Riverside, Ill , with instructions not to turn it over to us unless they know we are perfectly reliable and will do as we agree.

We have built up our business on these lines. We have merited a continuance of patronage from our customers. We have received the benefit of their words of commendation to their friends. We always do exactly as we agree

Our Guarantee

Perhaps there are many who feel that they are running some risk in ordering plans at a distance. We wish to assure our customers that there is no risk whatever. If, upon receipt of plans, you do not find them exactly as we represent them, if you do not find them complete and accurate in every respect, if you do not find them as well prepared as those furnished by any architect in the United States, or any that you have ever seen, we will refund your money upon the return of the plans from you in perfect condition.

All of our plans are prepared by licensed architects standing at the head of their profession, and the standard of their work is the very highest.

We could not afford to make this guarantee if we were not positive that we were furnishing the best plans put out in this country, even though our price is not more than one-seventh to one-tenth of the price usually charged.

Lumber Bill

We do not furnish a lumber bill. We state this here particularly, as some people have an idea that a lumber bill should accompany each

set of plans and specifications. In the first place, our plans are gotten up in a very comprehensive manner, so that any carpenter can easily take off the lumber bill without any difficulty. We realize that there are hardly two sections of the country where exactly the same kinds of materials are used, and, moreover, a lumber bill which we might furnish would not be applicable in all sections of the country. We furnish plans and specifications for houses which are built as far north as the Hudson Bay and as far south as the Gulf of Mexico. They are built upon the Atlantic and Pacific coasts, and you can also find them in Australia and South Africa. Each country and section of a country has its peculiarities as to sizes and qualities; therefore it would be useless for us to make a list that would not be universal. Our houses when completed may look the same, whether they are built in Canada or in Florida, but the same materials will not be used, for the reason that the customs of the people and the climatic conditions will dictate the kind and amount of materials to be used in their construction.

Estimated Cost

It is impossible for any one to estimate the cost of a building and have the figures hold good in all sections of the country. We do not claim to be able to do it. The estimated cost of the houses we illustrate is based on the most favorable conditions in all respects, and does not include plumbing and heating. We do not know your local conditions, and should we claim to know the exact cost of a building in your locality, a child would know that our statement was false. We advise consultation with your local responsible material dealers and reliable contractors, for they, and they alone, know your local conditions. We wish to be frank with you, and therefore make no statement that we cannot substantiate in every respect. If any plan in this book, or in any other book we publish, pleases you; if the arrangement of the rooms is satisfactory, and if the exterior is pleasing and attractive, then we make this claim—that it can be built as cheaply as if any other architect designed it, and we believe cheaper. We have studied economy in construction, and our knowledge of all the material that goes into the house qualifies us to give you the best for your money. We give you a plan that pleases you, one that is attractive, and one where every foot of space is utilized at the least possible cost. Can any architect do more, even at seven to ten times the price we charge you for plans?

Reversing Plans

We receive many requests from our patrons for plans exactly according to the designs illustrated, with the one exception of having them reversed or faced in the opposite direction. It is impossible for us to make this change and draw new plans, except at a cost of about eight times our regular prices. We see no reason why our regular plans will not answer your purpose. Your carpenter can face the house exactly as you wish it, and the plans will work out as well facing in one direction as in another. We can, however, if you wish, and so instruct us, make you a reversed blue print and furnish it at our regular price, but in that case all the figures and letters will be reversed, and therefore liable to cause as much confusion as if your carpenter reversed the plan himself while constructing the house. We would advise, however, in all cases where the plan is to be  reversed and there is the least doubt about the contractor not being able to work from the plans as we have them, that two sets of blue prints be purchased, one regular and the other reversed, and in such cases we will furnish two sets of blue prints and one set of specifications for only fifty per cent. added to our regular cost, making the $5.00 plan cost only $7.50.

Special Department

We have established a special department under the supervision of a licensed architect, to handle all special plans which our patrons may like to have drawn. We realize that often some special or original idea is wished carried out, and to provide for this we have our architects and draughtsmen. The price we charge is very reasonable. Should you wish the services of this department, it would be necessary for you to send us as full and complete information as possible, accompanied by a rough sketch illustrating as near as you can your ideas and requirements. Immediately upon receipt of this information from you, we will make you a price on these plans and specifications carrying out your own ideas, and if our price proves to be satisfactory, we will submit pencil sketch subject to your corrections and additions before proceeding to complete the plans. We must, however, have an understanding that we are under contract to do this work, for we cannot afford to do all the preliminary work without some guarantee that it will be accepted after we have agreed to make the plans entirely satisfactory to you. We will, however, make estimate on the cost of any special work, so that you will know exactly what it will cost you before we proceed with the plans.

How to Send Money

Remittances can be made by Post Office Money Order, Express Money Order, Bank Draft, United States or Canadian Bills. Take great care to write your address plainly, and be sure and write your name and address on the upper left-hand corner of the envelope. In addition, write plainly in your letter your name and address, the name of your city, county and State, if you are a resident of the United States, or if a resident of any other country, the name of the county, district or Province; also the street and number when necessary. We receive a great amount of money which it is impossible for us to trace on account of the incomplete or indistinct writing of name or address, and oftentimes the entire omission of both.

No. 1104

PRICE
of Plans and
Specifications
$5.00

House Design No. 1104

Full and complete working plans and specifications of this house will be furnished for $5.00. Cost of this house is from $1,150 to $1,300 according to the locality in which it is built.

FLOOR PLANS OF DESIGN No. 1104

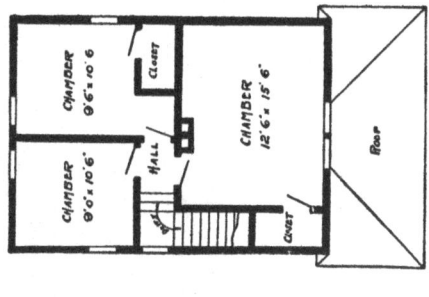

SECOND FLOOR PLAN

SIZE
Width, 22 feet.
Length, 31 feet

FIRST FLOOR PLAN

Blue prints consist of foundation plan; first and second floor plans; front, rear, two side elevations; wall sections and all necessary interior details. Specifications consist of about fifteen pages of typewritten matter.

No. 1085

PRICE of Plans and Specifications **$5.00**

House Design No. 1085

Full and complete working plans and specifications of this house will be furnished for $5.00. Cost of this house is from $1,450 to $1,600 according to the locality in which it is built.

FLOOR PLANS OF DESIGN No. 1085

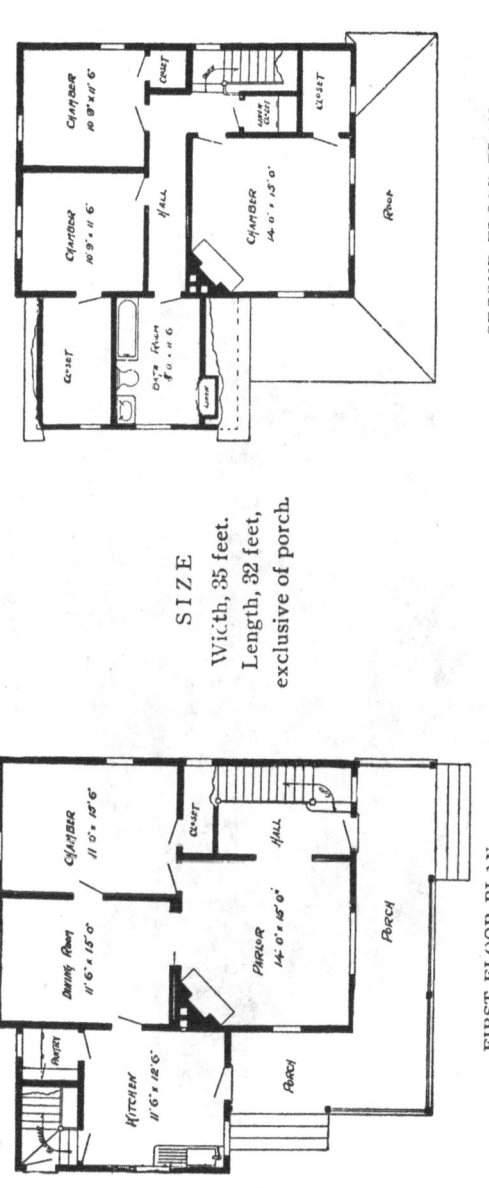

SECOND FLOOR PLAN

SIZE
Width, 35 feet.
Length, 32 feet, exclusive of porch.

FIRST FLOOR PLAN

Blue prints consist of cellar and foundation plan; first and second floor plans; front, rear, two side elevations; wall sections and all necessary interior details.

Specifications consist of about twenty pages of typewritten matter.

No. 1082

PRICE of Plans and Specifications **$5.00**

HOUSE DESIGN NO. 1082

Full and complete working plans and specifications of this house will be furnished for $5.00. Cost of this house is from $2,000 to $2,250 according to the locality in which it is built.

FLOOR PLANS OF DESIGN No. 1082

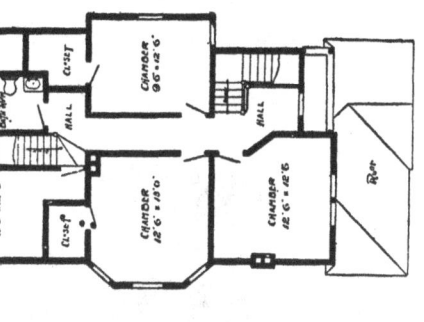

SECOND FLOOR PLAN

SIZE
Width, 28 feet.
Length, 45 feet,
exclusive of porch.

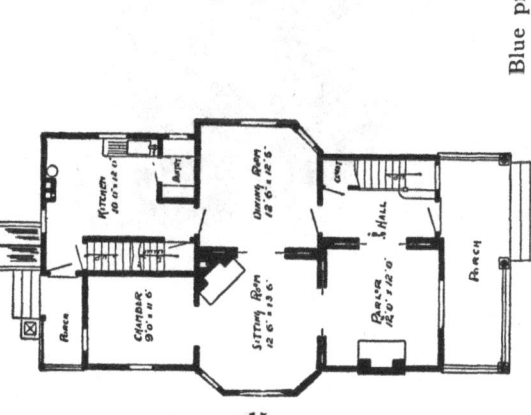

FIRST FLOOR PLAN

Blue prints consist of cellar and foundation plan; floor plans; front, rear, two side elevations; wall sections and all necessary interior details. Specifications consist of about twenty pages of typewritten matter.

No. 1002

PRICE
of Plans and
Specifications
$5.00

House Design No. 1002

Full and complete working plans and specifications of this house will be furnished for $5.00. Cost of this house is from $800 to $950 according to the locality in which it is built.

FLOOR PLANS OF DESIGN No. 1002

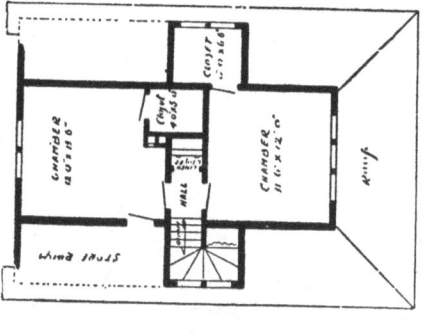

SECOND FLOOR PLAN

SIZE
Width, 24 feet.
Length, 30 feet, exclusive of porch.

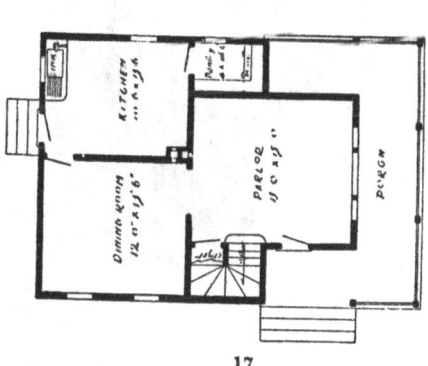

FIRST FLOOR PLAN

Blue prints consist of cellar and foundation plan; first and second floor plans; front, rear, two side elevations; wall sections and all necessary interior details.

Specifications consist of about fifteen pages of typewritten matter.

No. 1081

PRICE of Plans and Specifications **$5.00**

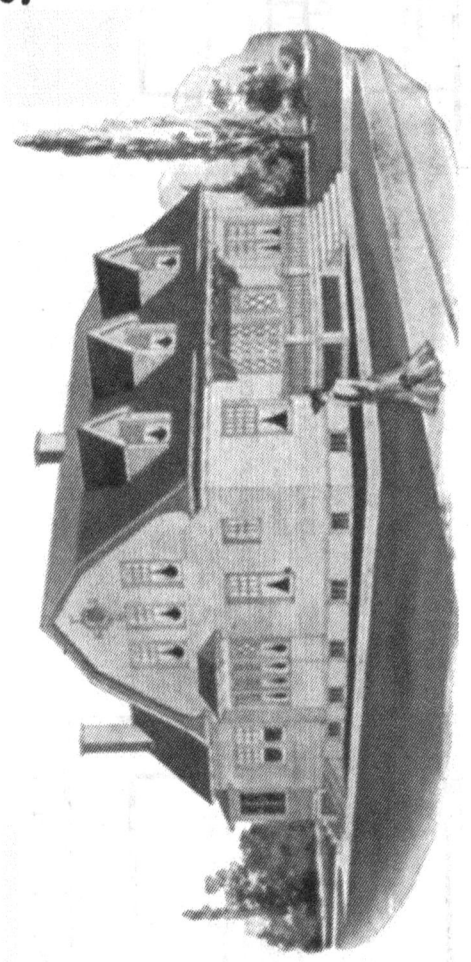

House Design No. 1081

Full and complete working plans and specifications of this house will be furnished for $5.00. Cost of this house is from $2,000 to $2,200 according to the locality in which it is built.

FLOOR PLAN OF DESIGN No. 1081

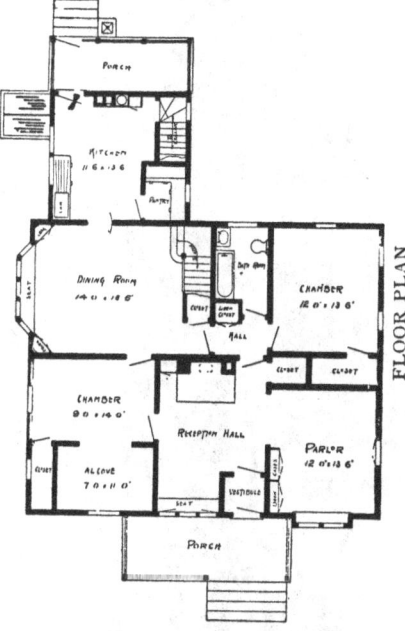

FLOOR PLAN

S I Z E

Width, 40 feet.   Length, 44 feet, exclusive of porches.

Blue prints consist of cellar foundation plan; floor plan; front, rear, two side elevations; wall sections and all necessary interior details.

Specifications consist of about fifteen pages of typewritten matter.

No. 1080

PRICE of Plans and Specifications $5.00

House Design No. 1080

Full and complete working plans and specifications of this house will be furnished for $5.00. Cost of this house is from $1,750 to $1,900 according to the locality in which it is built.

FLOOR PLANS OF DESIGN No. 1080

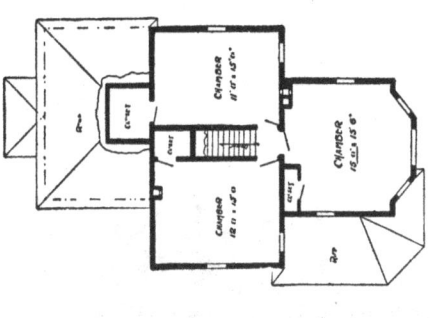

SECOND FLOOR PLAN

SIZE

Width, 28 feet.
Length, 45 feet,
exclusive of porches.

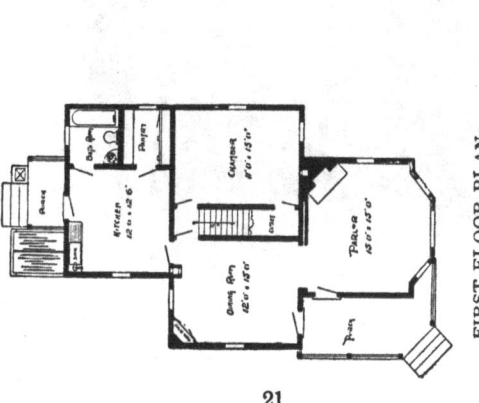

FIRST FLOOR PLAN

Blue prints consist of cellar and foundation plan; first and second floor plans; front, rear, two side elevations; wall sections and all necessary interior details.

Specifications consist of about twenty pages of typewritten matter.

No. 1075

PRICE
of Plans and
Specifications
$5.00

House Design No. 1075

Full and complete working plans and specifications of this house will be furnished for $5.00. Cost of this house is from $1,450 to $1,600 according to the locality in which it is built.

FLOOR PLAN OF DESIGN No. 1075

S I Z E
Width, 35 feet.
Length, 59 feet,
exclusive of porch.

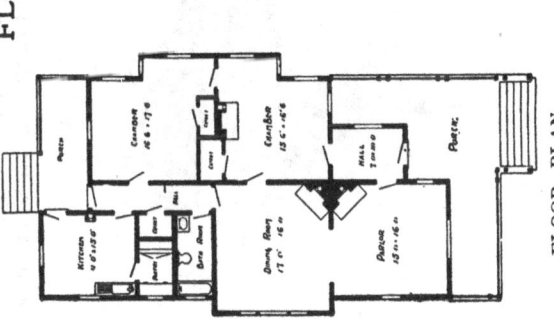

FLOOR PLAN

Blue prints consist of foundation plan; floor plan; front, rear, two side elevations wall sections and all necessary interior details.

Specifications consist of about fifteen pages of typewritten matter.

No. 1069

PRICE of Plans and Specifications **$5.00**

House Design No. 1069

Full and complete working plans and specifications of this house will be furnished for $5.00. Cost of this house is from $650 to $800 according to the locality in which it is built.

FLOOR PLAN OF DESIGN No. 1069

SIZE
Width, 33 feet, 6 inches.
Length, 26 feet.

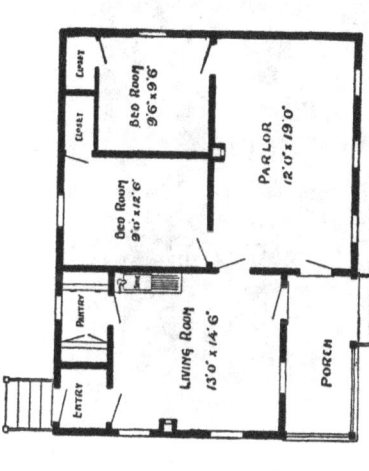

FLOOR PLAN

Blue prints consist of foundation plan; floor plan; front, rear, two side elevations; wall sections and all necessary interior details.
Specifications consist of about fifteen pages of typewritten matter.

No. 1067

PRICE
of Plans and Specifications
$5.00

House Design No. 1067

Full and complete working plans and specifications of this house will be furnished for $5.00. Cost of this house is from $1,050 to $1,200 according to the locality in which it is built.

FLOOR PLANS OF DESIGN No. 1067

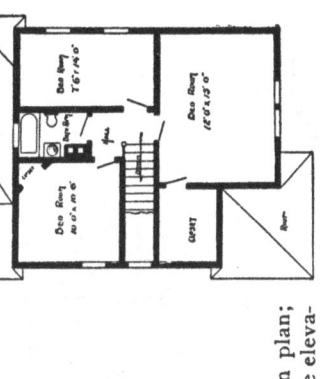

SECOND FLOOR PLAN

SIZE
Width, 24 feet.
Length, 40 feet,
exclusive of porch.

Blue prints consist of cellar and foundation plan; first and second floor plans; front, rear, two side elevations; wall sections and all necessary interior details.

Specifications consist of about fifteen pages of typewritten matter.

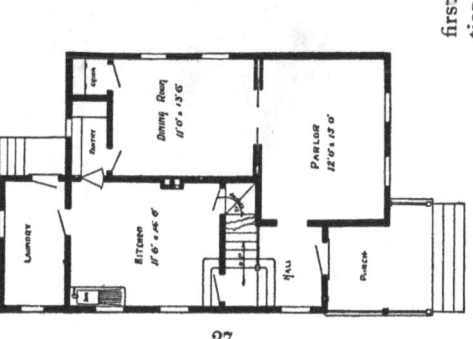

FIRST FLOOR PLAN

No. 1066

PRICE of Plans and Specifications **$5.00**

HOUSE DESIGN NO. 1066

Full and complete working plans and specifications of this house will be furnished for $5.00. Cost of this house is from $1,350 to $1,500 according to the locality in which it is built.

FLOOR PLANS OF DESIGN No. 1066

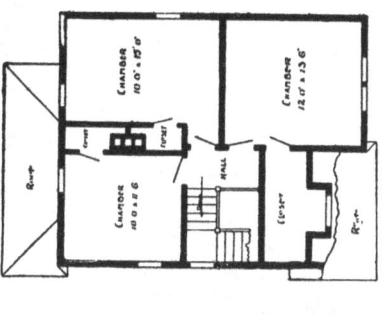

SECOND FLOOR PLAN

SIZE

Width, 24 feet,
Length, 30 feet,
exclusive of porch.

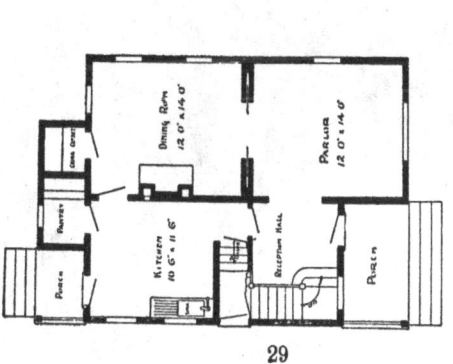

FIRST FLOOR PLAN

Blue prints consist of cellar and foundation plan; first and second floor plans; front, rear, two side elevations; wall sections and all necessary interior details.

Specifications consist of about twenty pages of typewritten matter.

No. 1064

PRICE
of Plans and
Specifications
$5.00

HOUSE DESIGN NO. 1064

Full and complete working plans and specifications of this house will be furnished for $5.00. Cost of this house is from $650 to $800 according to the locality in which it is built.

FLOOR PLAN OF DESIGN No. 1064

SIZE

Width, 22 feet. Length, 36 feet.

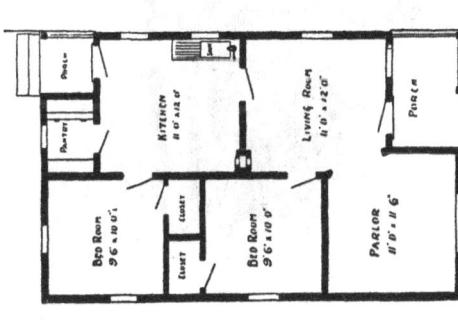

FLOOR PLAN

Blue prints consist of foundation plan; floor plan; front, rear, two side elevations; wall sections and all necessary interior details.

Specifications consist of about fifteen pages of typewritten matter.

No. 1023

PRICE of Plans and Specifications **$5.00**

HOUSE DESIGN No. 1023

Full and complete working plans and specifications of this house will be furnished for $5.00. Cost of this house is from $2,000 to $2,250 according to the locality in which it is built.

FLOOR PLANS OF DESIGN No. 1023.

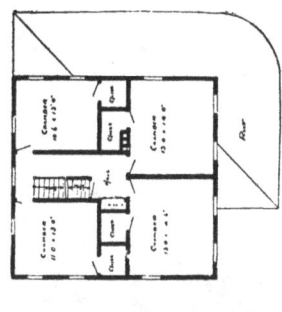

SECOND FLOOR PLAN

SIZE
Width, 30 feet.
Length, 32 feet,
exclusive of porches.

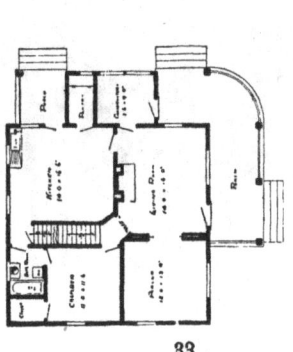

FIRST FLOOR PLAN

Blue prints consist of cellar and foundation plan; first and second floor plans; front, rear, two side elevations; wall sections and all necessary interior details.

Specifications consist of about twenty pages of typewritten matter.

No. 1099

PRICE of Plans and Specifications
$5.00

House Design No. 1099

Full and complete working plans and specifications of this house will be furnished for $5.00. Cost of this house is from $1,650 to $1,800 according to the locality in which it is built.

FLOOR PLAN OF DESIGN No. 1099

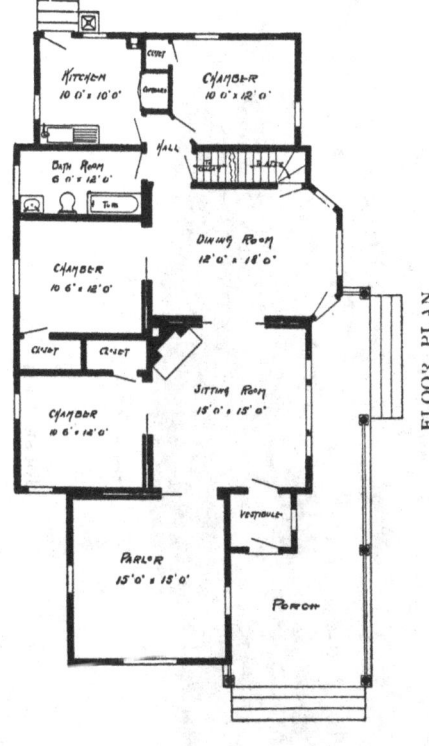

FLOOR PLAN

SIZE.

Width, 32 feet. Length, 59 feet.

Blue prints consist of cellar and foundation plan; floor plan; front, rear, two side elevations; wall sections and all necessary interior details.

Specifications consist of about fifteen pages of typewritten matter.

No. 1047

PRICE of Plans and Specifications **$5.00**

House Design No. 1047

Full and complete working plans and specifications of this house will be furnished for $5.00. Cost of this house is from $1,400 to $1,550 according to the locality in which it is built.

FLOOR PLANS OF DESIGN No. 1047

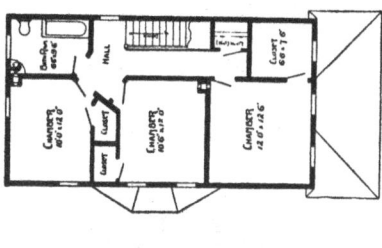

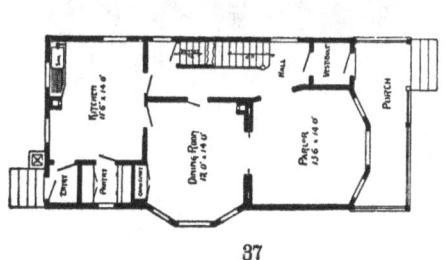

FIRST FLOOR PLAN

SECOND FLOOR PLAN

SIZE

Width, 22 feet.
Length, 40 feet,
exclusive of porch.

Blue prints consist of cellar and foundation plan; first and second floor plans; front, rear, two side elevations; wall sections and all necessary interior details.

Specifications consist of about twenty pages of typewritten matter.

No. 1087

PRICE of Plans and Specifications $8.00

House Design No. 1087

Full and complete working plans and specifications of this house will be furnished for $8.00. Cost of this house is from $2,750 to $2,900 according to the locality in which it is built

FLOOR PLANS OF DESIGN No. 1087

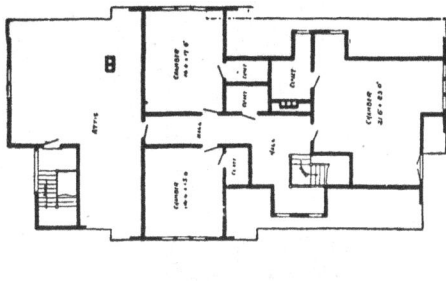

SECOND FLOOR PLAN

S I Z E

Width, 40 feet.
Length, 67 feet,
exclusive of porch.

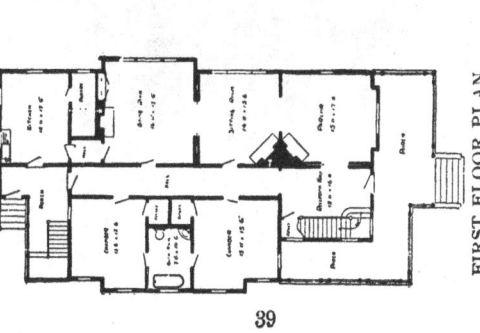

FIRST FLOOR PLAN

Blue prints consist of foundation plan; first and second floor plans; front, rear, two side elevations; wall sections and all necessary interior details.

Specifications consist of about twenty pages of typewritten matter.

No. 1044

PRICE
of Plans and
Specifications
$5.00

House Design No. 1044

Full and complete working plans and specifications of this house will be furnished for $5.00. Cost of this house is from $1,650 to $1,800 according to the locality in which it is built.

FLOOR PLANS OF DESIGN No. 1044

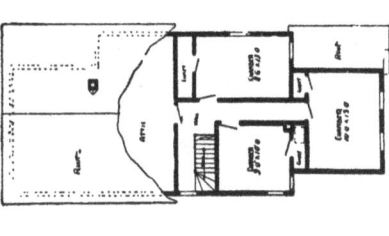

SECOND FLOOR PLAN

S I Z E
Width, 22 feet.
Length, 52 feet.

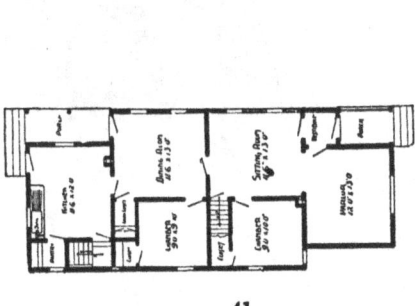

FIRST FLOOR PLAN

Blue prints consist of cellar and foundation plan; first and second floor plans; front, rear, two side elevations; wall sections and all necessary interior details.
Specifications consist of about twenty pages of typewritten matter.

No. 1105

PRICE
of Plans and Specifications
$5.00

House Design No. 1105

Full and complete working plans and specifications of this house will be furnished for $5.00. Cost of this house is from $1,550 to $1,700 according to the locality in which it is built.

FLOOR PLANS OF DESIGN No. 1105

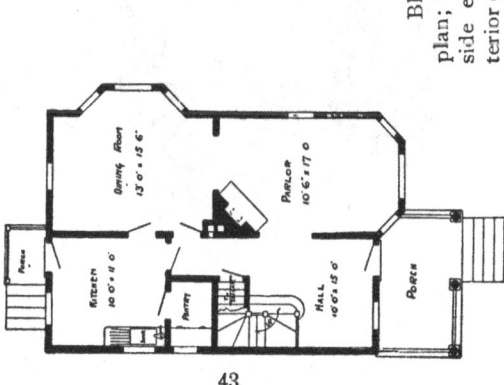

SECOND FLOOR PLAN

FIRST FLOOR PLAN

SIZE

Width, 25 feet.
Length, 34 feet, 6 inches, exclusive of porches.

Blue prints consist of cellar and foundation plan; first and second floor plans; front, rear, two side elevations; wall sections and all necessary interior details.

Specifications consist of about twenty pages of typewritten matter.

No. 1015

PRICE of Plans and Specifications **$7.50**

House Design No. 1015

Full and complete working plans and specifications of this house will be furnished for $7.50. Cost of this house is from $3,700 to $4,000 according to the locality in which it is built.

FLOOR PLANS OF DESIGN No. 1015

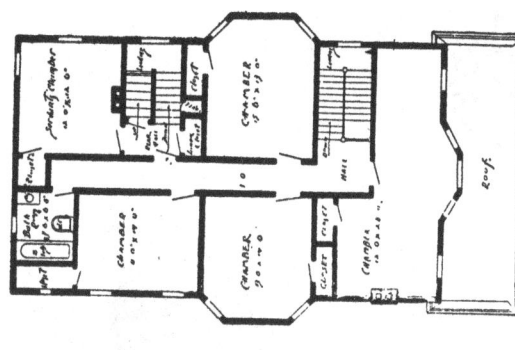

SECOND FLOOR PLAN

S I Z E
Width, 34 feet.
Length, 50 feet,
exclusive of porches.

Blue prints consist of cellar and foundation plan; first and second floor plans; front, rear, two side elevations; wall sections and all necessary interior details. Specifications consist of about twenty pages of typewritten matter.

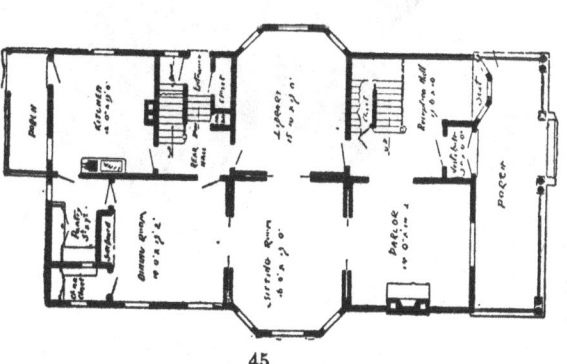

FIRST FLOOR PLAN

No. 1050

PRICE of Plans and Specifications
$5.00

House Design No. 1050

Full and complete working plans and specifications of this house will be furnished for $5.00. Cost of this house is from $650 to $800 according to the locality in which it is built

FLOOR PLANS OF DESIGN No. 1050

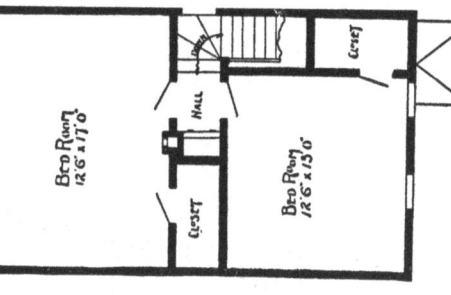

SECOND FLOOR PLAN

SIZE

Width, 18 feet.
Length, 30 feet,
exclusive of porch.

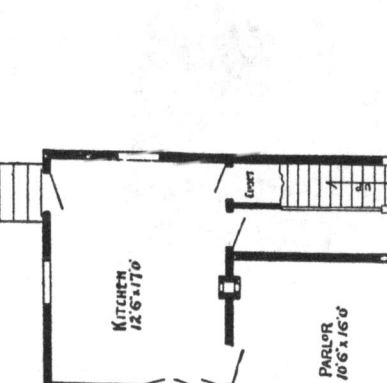

FIRST FLOOR PLAN

Blue prints consist of foundation plan; first and second floor plans; front, rear, two side elevations; wall sections and all necessary interior details. Specifications consist of about fifteen pages of typewritten matter.

No. 1068

PRICE
of Plans and
Specifications
$5.00

House Design No. 1068

Full and complete working plans and specifications of this house will be furnished for $5.00. Cost of this house is from $1,600 to $1,750 according to the locality in which it is built.

FLOOR PLANS OF DESIGN No. 1068

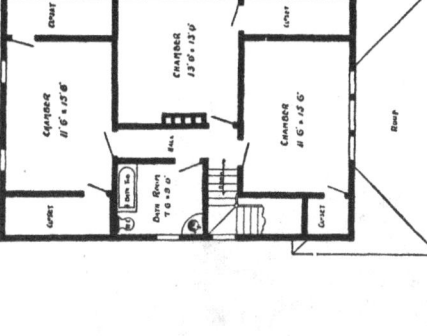

SECOND FLOOR PLAN

SIZE
Width, 27 feet.
Length, 38 feet,
exclusive of porches.

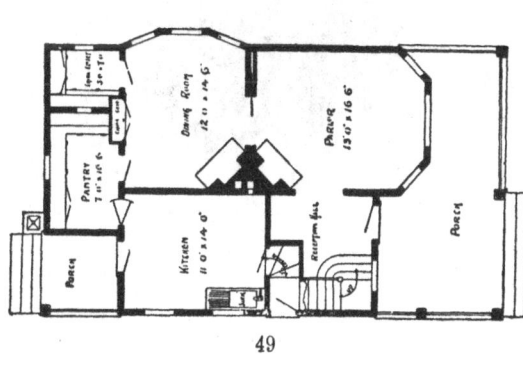

FIRST FLOOR PLAN

Blue prints consist of cellar and foundation plan; first and second floor plans; front, rear, two side elevations; wall sections and all necessary interior details. Specifications consist of about fifteen pages of typewritten matter.

No. 1062

PRICE of Plans and Specifications **$5.00**

House Design No. 1062

Full and complete working plans and specifications of this house will be furnished for $5.00. Cost of this house is from $1,650 to $1,800 according to the locality in which it is built.

FLOOR PLANS OF DESIGN No. 1062

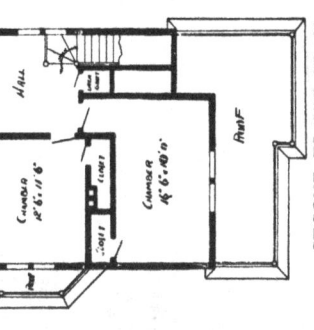

SECOND FLOOR PLAN

SIZE

Width, 27 feet.
Length, 41 feet,
exclusive of porch.

FIRST FLOOR PLAN

Blue prints consist of cellar and foundation plan; first and second floor plans; front rear, two side elevations; wall sections and all necessary interior details. Specifications consist of about twenty pages of typewritten matter.

No. 1102

PRICE of Plans and Specifications $5.00

House Design No. 1102

Full and complete working plans and specifications of this house will be furnished for $5.00. Cost of this house is from $1,550 to $1,700 according to the locality in which it is built.

FLOOR PLANS OF DESIGN No. 1102

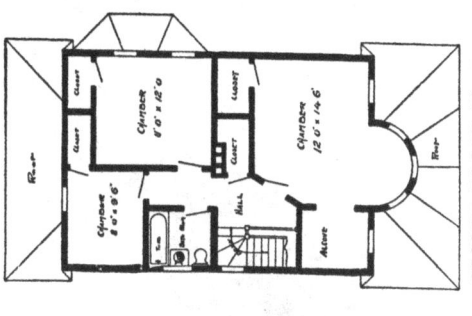

SECOND FLOOR PLAN

SIZE

Width, 24 feet, 6 inches.
Length, 37 feet,
exclusive of porches.

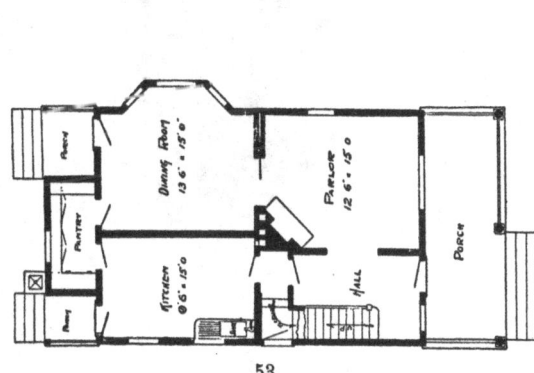

FIRST FLOOR PLAN

Blue prints consist of cellar and foundation plan; first and second floor plans; front, rear, two side elevations; wall sections and all necessary interior details. Specifications consist of about twenty pages of typewritten matter.

No. 1017

PRICE of Plans and Specifications **$5.00**

HOUSE DESIGN No. 1017

Full and complete working plans and specifications of this house will be furnished for $5.00. Cost of this house is from $1,450 to $1,600 according to the locality in which it is built.

FLOOR PLANS OF DESIGN No. 1017

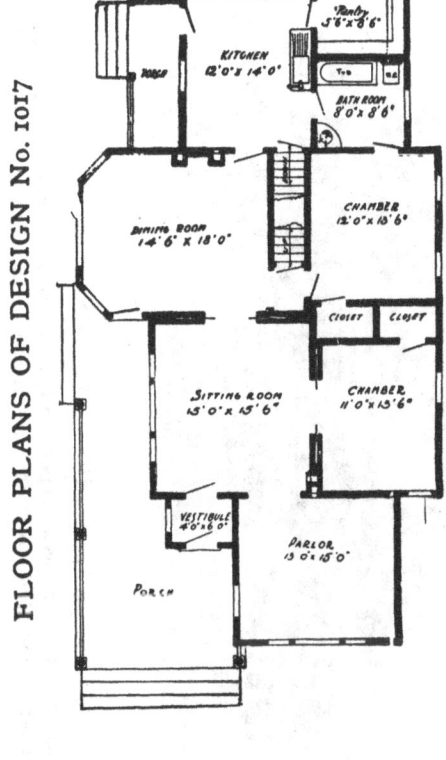

FLOOR PLAN

SIZE

Width, 34 feet.　　Length, 60 feet.

Blue prints consist of cellar and foundation plan; floor plan; roof plan; front, rear, two side elevations; wall sections and all necessary interior details.

Specifications consist of about fifteen pages of typewritten matter.

No. 1025

PRICE of Plans and Specifications **$5.00**

House Design No. 1025

Full and complete working plans and specifications of this house will be furnished for $5.00. Cost of this house is from $1,950 to $2,200 according to the locality in which it is built.

FLOOR PLANS OF DESIGN No. 1025

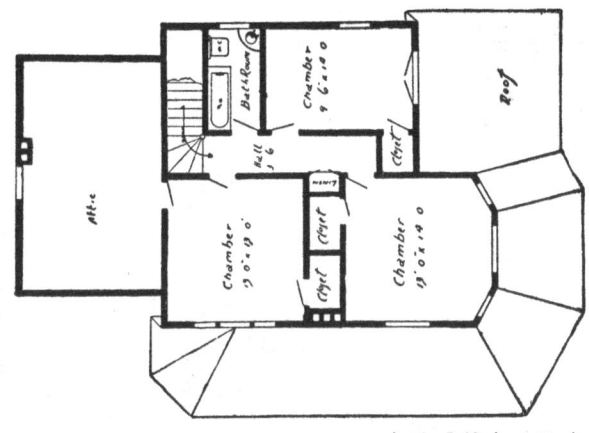

SECOND FLOOR PLAN

FIRST FLOOR PLAN

SIZE
Width, 35 feet.
Length, 46 feet, exclusive of porches.

Blue prints consist of cellar and foundation plan; first and second floor plans; front, rear, two side elevations; wall sections and all necessary interior details.

Specifications consist of about twenty pages of typewritten matter.

No. 1049

PRICE of Plans and Specifications **$5.00**

House Design No. 1049

Full and complete working plans and specifications of this house will be furnished for $5.00. Cost of this house is from $1,500 to $1,650 according to the locality in which it is built.

FLOOR PLAN OF DESIGN No. 1049

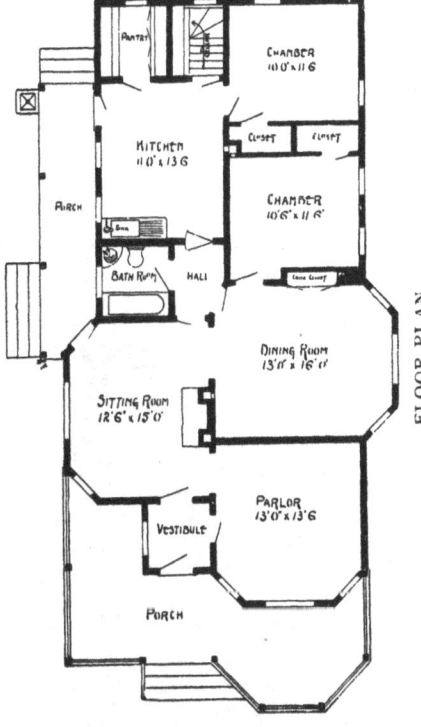

FLOOR PLAN

SIZE

Width, 30 feet. Length, 52 feet, exclusive of porches.

Blue prints consist of cellar foundation plan; floor and roof plans; front, rear, two side elevations; wall sections and all necessary interior details. Specifications consist of about fifteen pages of typewritten matter.

www.ingramcontent.com/pod-product-compliance
Lightning Source LLC
Chambersburg PA
CBHW011950150426
43195CB00018B/2879